First Published 2010
Reprinted August 2010

The Educational Company of Ireland
Ballymount Road
Walkinstown
Dublin 12

A member of the Smurfit Kappa Group plc

Artists: Maria Murray, Peter Donnelly
Design and Layout: Outburst Design
Cover Design: One House
Cover Images: Shutterstock

Acknowledgements
The author wants to thank Riona Browne, Karl Lynch and all the staff of Killeeneen NS for their assistance with this publication.

Photographs are courtesy of Alamy, Dreamstime, Fotolia, Getty, Pixelio, Science Photo Library, Shutterstock.

Further information on each chapter is provided in the Teacher's Resource Book.

The Educational Company of Ireland

Contents

Chapter	Topic	Strand	Strand Unit	Page
1	Bridge Building	Materials Environmental awareness and care Energy and forces	Properties and characteristics of materials Science and the environment Environmental awareness Caring for our environment Forces	4
2	Balancing the Load	Energy and forces Environmental awareness and care	Forces Science and the environment	9
3	Free Fall	Energy and forces Materials Environmental awareness and care	Forces Properties and characteristics of materials Science and the environment	13
4	Sleepy Heads	Living things Environmental awareness and care	Plant and animal life Science and the environment	18
5	Food for Life	Living things Materials Environmental awareness and care	Human life Properties and characteristics of materials Science and the environment	23
6	Hot and Cold	Energy and forces Environmental awareness and care	Heat Science and the environment	28
7	Soaking Up	Materials Environmental awareness and care	Properties and characteristics of materials Materials and change Science and the environment	33
8	Switch it on!	Energy and forces Environmental awareness and care	Electricity Science and the environment	39
9	Batteries and Bulbs	Energy and forces Environmental awareness and care	Electricity Science and the environment	45

Contents

Chapter	Topic	Strand	Strand Unit	Page
10	The Nature of Things	Materials Environmental awareness and care	Properties and characteristics of materials Science and the environment	50
11	A Cosy Nest	Living things Environmental awareness and care	Plant and animal life Science and the environment	55
12	Loud and Clear	Energy and forces Environmental awareness and care	Forces Science and the environment	61
13	Nature's Ploughman	Living things Environmental awareness and care	Plant and animal life Science and the environment	66
14	Say Cheese!	Living things Environmental awareness and care	Plant and animal life Science and the environment	71
15	Water Wizardry	Materials Environmental awareness and care	Properties and characteristics of materials Materials and change Science and the environment	77
16	Flower Power	Living things Environmental awareness and care	Plant and animal life Science and the environment	82
17	Moving Air	Materials Environmental awareness and care Energy and forces	Properties and characteristics of materials Materials and change Science and the environment Forces	89
18	Plants Provide!	Living things Environmental awareness and care	Human life Plant and animal life Science and the environment	93

Bridge Building

How will I cross this river?

The best way to cross the river would be to swim.

I would build a bridge over it.

I would use a boat to cross it.

I would jump across it.

Identify the main question you need to answer.
Record it now on your Investigation Record Sheet.

Use your Record Sheet to investigate!
- My Question
- My Prediction
- Let's Investigate!
- My Record
- My Results
- My Conclusion

My Prediction

Complete this sentence on your Investigation Record Sheet:

I think that the best way to make the bridge with paper or card is to ______________________________

Let's Investigate!

You will need:
- A4 sheets of paper
- A4 sheets of card
- Some small books

1 Get an A4 sheet of paper.
Can you make it stand up?
What did you do to make it stand?

2 Now that the paper is standing, would it support another page?
How many pages will it support?

3 Take the page and fold it in a different way.
Can you make it stronger? How?
Find out if it will support a small book.

4 Try out other ways of folding the page to make it stronger. How many small books can it support?

5 Get a sheet of card.
Fold it in such a way that it will be very strong.
How many small books will it support?

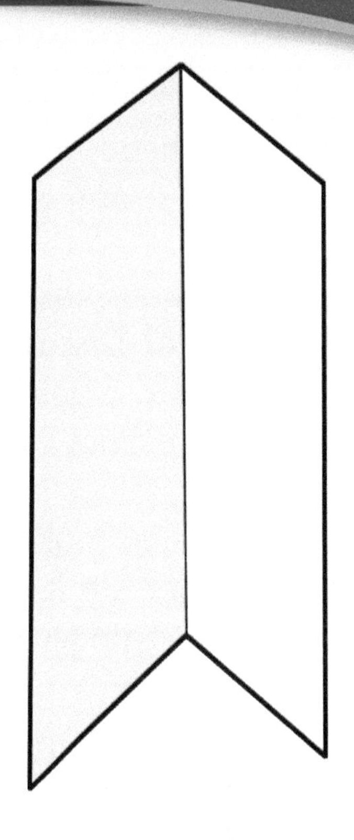

On your Investigation Record Sheet, draw diagrams of your investigations.

It is now time to record your results.

Now that your class has discussed the investigations, you can record the conclusion.

You will need:
- A4 sheets of paper
- A4 sheets of card
- Some small books

Finding Out More

Let's make a single-span bridge over a small river, like the one in the diagram.

1 With your group, discuss what you will use for the sides of the bridge.
What will you use for the bridge itself?
Will you use paper or cardboard? How wide will you make the river? How will you test the strength of the bridge?

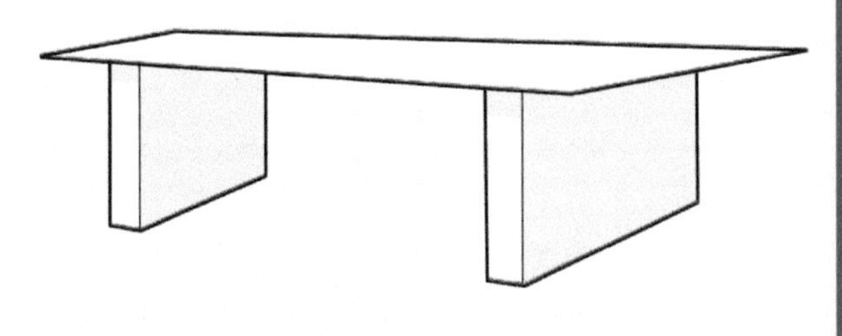

2 Go ahead and make the bridge as you planned. How strong is it?

3 Can you make some changes that will strengthen the bridge? Discuss this with your group.
What did you decide? Make the bridge stronger and find out how strong it is.

4 Can you make one more change to improve the bridge? Do it and find out if you have now made an even stronger bridge.

My Record

On your Investigation Record Sheet, draw diagrams of your investigations.

My Results

It is now time to record your results.

My Conclusion

Now that your class has discussed the investigations, you can record the conclusions.

Detective Time

1 Find out about each of these famous bridges:
(a) The Tower Bridge in London
(b) The Golden Gate Bridge in San Francisco
(c) Sydney Harbour Bridge.

For each bridge, find out:
- Who designed it?
- When was it built?
- Why was it built?
- What materials were used to make it?
- How wide is the **span** of the bridge?

span: the distance between two supports or points

2 How many different types of bridge are there?
In your group compile a Powerpoint presentation for your class on one type of bridge. Other groups could choose other types of bridge and make their own presentations.

3 Is there a bridge near your school?
Do some research:
- What type of bridge is it?
- Why is it there?
- Who built it?
- When was it built?
- What materials were used to make it?
- Who looks after the bridge?

My Record

On your Investigation Record Sheet, draw diagrams of a bridge near your school.

My Results

It is now time to record the answers to the questions asked.

Very Interesting

Science and the environment

- Scientists and engineers have used a variety of materials to build bridges. They have used wood, stone, concrete, iron and steel. Steel is a strong mix of iron and elements like carbon. Scientists and engineers often mix materials to make them better for the job. Concrete can be made stronger by positioning steel bars within it. This concrete and steel mix is often used to make bridges. Can you think of any other mixes that improve materials?

Web Watch!
If you would like to find out more about early bridges, go to this website:
http://www.factmonster.com/ce6/sci/A0857018.html

Integration Project

English

An idiom is a phrase where the words together have a different meaning from the dictionary definitions of the individual words. 'Don't burn your bridges' is an idiom. Can you figure out what is meant by this? Give an example of where it could be used.

Gaeilge

The Irish word for bridge is 'Droichead'. Many Irish place names have the word Droichead, e.g. An Droichead Nua (Newbridge). An féidir leat ainm aon áit eile a fháil leis an bhfocal 'Droichead' ann?

Geography

1. Choose a county in Ireland and list 3 rivers in that county. Find out the names of any bridges crossing these rivers.
2. Name a famous bridge in each of these places: Ireland, England, America, Australia.
3. Find out where in the world you would see the following:
   - The tallest bridge
   - The widest bridge
   - The longest bridge.

Mathematics

The span of a bridge is the length of a bridge. It is said that the length of your arm span is approximately the same as your height. Take measurements of the arm spans and the heights of each other and check out if this theory is true.

History

Find out the history behind the song 'London Bridge is Falling Down'.

http://www.talewins.com/world travel/londonbridge.htm

You can find the lyrics and music of the song on:

http://kids.niehs.nih.gov/lyrics/london.htm

Science

Do further research on bridges and discover some different types of bridge.

http://www.pbs.org/wgbh/buildingbig/bridge/index.html

http://www.hevanet.com/bridgink/

Visual Arts

Design and draw a bridge.

Where is the centre of gravity in the pencil?

The centre of gravity is in the middle.

The centre of gravity could be at the tip.

I predict that there are a few centres of gravity.

A pencil is too small to have a centre of gravity.

The BIG Question

Identify the main question you need to answer.
Record it now on your Investigation Record Sheet.

Use your Record Sheet to investigate!
My Question
My Prediction
Let's Investigate!
My Record
My Results
My Conclusion

My Prediction

Complete this sentence on your Investigation Record Sheet:

I think that the centre of gravity is ____________________
____

Let's Investigate!

1 Begin by locating a ruler. Place it on your finger as in the diagram. Move it about until you find the point at which it remains **level and balanced**.

You will need:
- A ruler
- A book
- A pencil
- Lunch box
- Selection of items

You have now found the **centre of gravity** of the ruler.
How far from the end is the centre of gravity?

2 Does the ruler have more than one centre of gravity?
Try to balance it in other positions.

3 Gather together a number of items. Get a book, a pencil, your lunch box and one or two other items.
Try to find the centre of gravity of each of them.

My Record

On your Investigation Record Sheet, draw diagrams of your investigations.

My Results

It is now time to record your results.

My Conclusion

Now that your class has discussed the investigations, you can record the conclusions.

Finding Out More

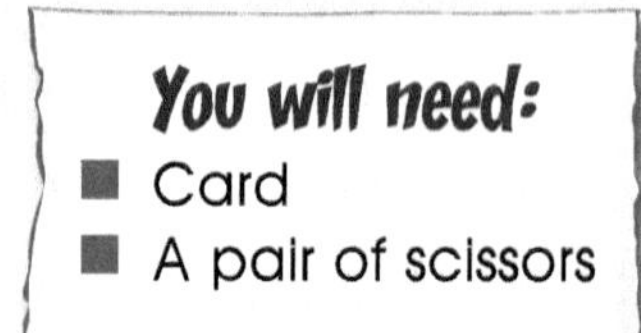

1 Get a piece of cardboard.
Draw an irregular shape on it.
Get a scissors and cut out the shape.

2 Mark the spot where you think you will find the centre of gravity of this shape. Try it out. Were you correct?

3 What did you notice about the position of the centre of gravity of the irregular shape?

4 Make a few more irregular shapes and see if you can find their centres of gravity.

Record your diagrams, results and conclusion on your Investigation Record Sheet, just as you did for the Let's Investigate! exercises.

Detective Time

You will need:
- A ruler
- A pencil
- Plasticine
- Two 10c coins
- Two 50c coins

1 Place a pencil on the table. Put a ruler on it and move the ruler about until you find its centre of gravity. You may have to put some plasticine on the pencil and the ruler to secure their positions.

2 Get a 10 cent coin and place it on the ruler. Position it so that it is centred on the 5cm mark. What happens to the ruler?

3 Discuss with your group where you might place another 10 cent coin, so that you could balance the ruler again. Try it out. What did you find?

4 Repeat the experiment but this time use two 50 cent coins. Place the first 50 cent coin at the position on the ruler that reads 12cm. Where, do you think, will the other 50 cent coin need to be placed, so that the ruler will be balanced again? Try it out. What do you notice?

My Record

On your Investigation Record Sheet, draw diagrams of your investigations.

My Results

It is now time to record your results.

My Conclusion

Now that your class has discussed the investigations, you can record the conclusions.

Make and Do

Design and make a see-saw

- **Look around and see what materials are available.**
- **Work with your group to draw a design of a see-saw.**
- **Show the plans to your teacher.**
- **Use some available materials to make a see-saw.**
- **Draw a diagram of the finished product.**

Very Interesting

Science and the environment

- We use levers almost every day to help us to do work: nutcrackers, bottle openers, the hinges on our doors, scissors, tweezers and nail clippers. Can you think of others?

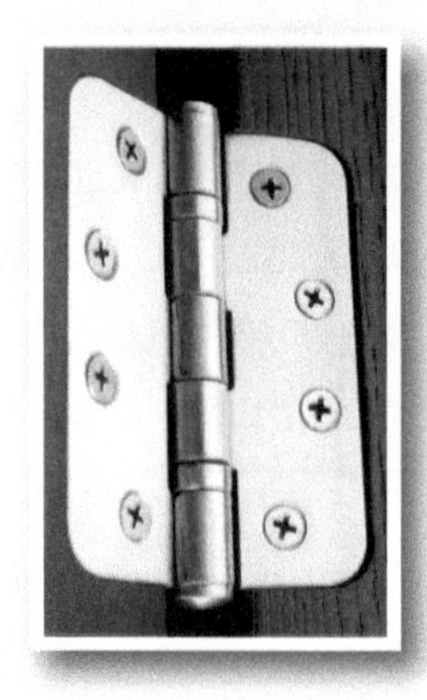

Web Watch!
If you would like to find out more about levers and machines, go to this website:
http://www.enchantedlearning.com/physics/machines/Levers.shtml

Balancing the Load

English

An alien has landed in your house. As he has never seen these household gadgets before, write out instructions for him in the use of one of them – scissors, tweezers, tongs, nutcracker or stapler.

Gaeilge

Ainmnigh na baill beatha a úsáideann 'levers'.

Mathematics

Using a scissors, make right angles, acute angles and obtuse angles. Now make these angles using your own body parts.

History

Levers have been used by people for thousands of years. Find out more on: http://www.historyforkids.org/scienceforkids/physics/machines/lever.htm

SPHE

Levers make life easier by helping us to lift things.
List other ways in which life can be made easier for us.

Geography

List as many levers as you can that you see in the world around you.

Science

Can you find levers in your own body? For more information visit: http://www.dynamicscience.com.au/tester/solutions/hydraulicus/humanbody.htm

Free Fall 3

Which object will hit the ground first?

My sheet of paper will fall slowly because the air will hold it up.

My orange will fall very quickly because it is heavy.

My paper clip will fall slowly because it is small.

The book will fall slowly because it is wide and flat.

The BiG Question

Identify the main question you need to answer. Record it now on your Investigation Record Sheet.

Use your Record Sheet to investigate!
My Question
My Prediction
Let's Investigate!
My Record
My Results
My Conclusion

My Prediction

Complete this sentence on your Investigation Record Sheet:

I think that the ______________________________

____

Let's Investigate!

You will need:
- An orange
- A grape

1. Begin by discussing how your group is going to do this experiment safely.
   - How will you get the grape and the orange up high?
   - Who will hold them?
   - Will you use two people to hold them or would one person be able to hold both?
   - How will you make sure that both pieces of fruit are dropped at exactly the same time?
   - How will you decide which one hits the ground first?

2. Tell your teacher how you plan to do the investigation.
3. Once your teacher is satisfied with your plan, do the investigation.
4. It is a good idea to do it a few times to make sure that your conclusions are correct.

My Record

On your Investigation Record Sheet, draw diagrams of your investigations.

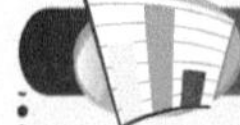

My Results

It is now time to record your results.

My Conclusion

Now that your class has discussed the investigations, you can record the conclusion.

Finding Out More

You will need:
- A 5 cent coin
- A €1 coin
- A small stone
- A small marble
- A variety of objects

There are three investigations to be done. Read them and then discuss them with your group.

In your group:
- Decide how you will do each investigation.
- Decide how you will record the results.
- Gather the materials needed and do the investigation.

1. Get a 5 cent coin and a €1 coin. Drop both from the same height and observe which one hits the ground first. What did you find?

2. Get a small stone about the size of your fist and also a small marble. Drop both from the same height and watch to see which one hits the ground first. What did you find?

3. Choose a few more objects of different weights and sizes and repeat the experiments with them. What did you find? Draw some conclusions.

My Record

On your Investigation Record Sheet, draw diagrams of your investigations.

My Results

It is now time to record your results.

My Conclusion

Now that your class has discussed the investigations, you can record the conclusion.

Detective Time

You will need:
- An A4 sheet of paper
- Scissors

1. **Which will fall faster?**
   - Divide an A4 sheet of paper in two.
   - Crumple one half and leave the other half as it is.
   - We are going to find out which piece of paper will fall faster than the other.

   In your group:
   - Decide how you will design an investigation to answer the big question!
   - Decide how you will record the results.
   - Gather the materials needed and do the investigation.

2 A parachute

- What is a parachute?
- List some of the uses of a parachute.
- How does a parachute work?

Make and Do

Design and make a small parachute which will be able to land a small marble. Draw up a plan. Discuss it with your teacher. Now make the parachute.

Record your diagrams, results and conclusion on your Investigation Record Sheet, just as you did for the Let's Investigate! exercises.

Very Interesting

Science and the environment

- Scientists have experimented with different types of parachute for centuries. In 1911 Grant Morton made the first parachute jump from an aeroplane. The following year, a parachute was used to drop a large parcel.
- Nowadays, parachutes are very well designed, with many safety features. Most are made from nylon because this material is light and very strong.
- Parachutes have many uses:
  - They are used to slow down the fall of an object.
  - They are used to slow down the speed of an aeroplane that is landing on a large aircraft carrier.
  - They are used in times of disaster or in times of famine, when there is a need to get food quickly to a remote area. Aeroplanes carry food packs over the area and these are dropped to the ground, using parachutes.

Web Watch!
If you would like to find out more, go to this website:
http://www.light-science.com/newtonapple.html

Integration Project

English

Use the following questions (Who, What, When, Where and Why) to write a non-rhyming poem about leaves falling from the trees – a 5W poem. Here is a 5W poem about autumn leaves:

Autumn leaves *(Who or what)*
Falling from the trees *(What action)*
Towards the end of the year *(When)*
In my garden *(Where)*
Because they have died *(Why).*

Gaeilge

1 Bhí Úna ag dul ar scoil maidin amháin. Go tobann, thit sí den rothar ... Críochnaigh an scéal.
2 Tomhais: Cén rud a bhíonn ag fás síos?

Mathematics

1 Falling Prices. In the autumn sales the prices of all items of clothing have fallen by 1/4 of their original price. If jeans originally cost €20, hoodies cost €16 and runners cost €24, how much are these items costing now?
2 Using an egg timer, see how fast you can complete a sheet of tables.

Free Fall

SPHE

Chicken Licken is an old story about a chicken who believes that the sky is falling.
Read the story and answer this question: What, do you think, is the moral in this story?

History

In your own locality, find and investigate an old ruin that has fallen down.

Music

Create a rainstorm with rain sticks.

Geography

1 Measure the rainfall every day for the next month and record your findings.
2 Compare the rainfall of two different months during the year.
3 Investigate famous waterfalls in the world.

Science

Investigate the famous scientist Sir Isaac Newton on:

http://www.light-science.com/newtonapple.html

4 Sleepy Heads

Why do some creatures hibernate?

They sleep in winter because it is very cold.

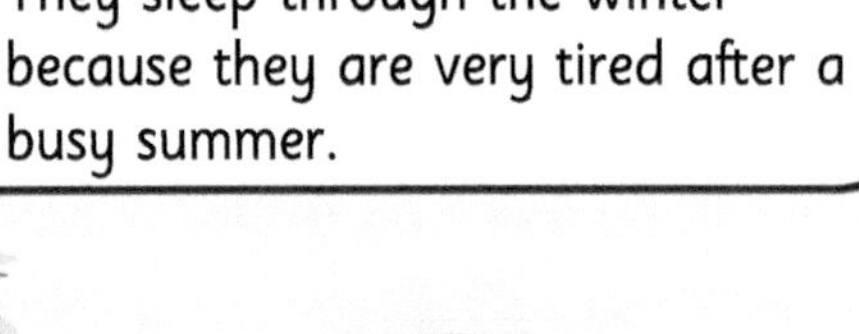

They sleep through the winter because they are very tired after a busy summer.

They sleep because there isn't any food in the winter.

They sleep because they are lazy.

The BIG Question

Identify the main question you need to answer. Record it now on your Investigation Record Sheet.

to hibernate: to sleep through the winter months

My Prediction

Complete this sentence on your Investigation Record Sheet:

I think that the hedgehog hibernates because ___________

Let's Investigate!

During the winter, all creatures have to overcome two big problems: how to keep warm and how to find food. When creatures hibernate, they seem to be able to cope with these problems.

The hedgehog

Find out how the hedgehog survives in winter.

1 How does a hedgehog survive without food for the winter?

2 How does the hedgehog keep warm in winter?

3 Where does the hedgehog make its home for the winter?

4 Find out as much information as you can about the hedgehog. Write up a short project and present it to the class.

My Record

Draw a mind map with the information that you have gathered on the hedgehog.

My Conclusion

Now that your class has discussed the information that you have on the hedgehog, answer this question: Why does the hedgehog hibernate?

Finding Out More

The frog

1 Frogs also hibernate during the winter.

2 Find out all that you can about frogs.
   (a) Begin by learning about the life cycle of the frog.
   (b) A frog is an amphibian. What does that mean?
   (c) A frog is also cold-blooded. What does that mean?
   (d) What do frogs eat?
   (e) A frog's tongue is unusual. Why?
   (f) How does a frog camouflage itself?
   (g) Where does a frog hibernate?

3 Do a short project on 'The Frog'.

My Record

Draw a mind map on the information that you have on the frog.

My Conclusion

Now that your class has discussed frogs, answer this question: Why does the frog hibernate?

Detective Time

The bat

1 Bats hibernate in winter.

2 Let's find out all we can about bats.
   (a) A bat is the only mammal that can fly. What is a mammal?
   (b) Describe the appearance of a bat.
   (c) Bats are nocturnal creatures. What does that mean?
   (d) In addition to their eyes, what else do bats use to help them find their way, as they fly and hunt?
   (e) What is the largest bat in the world?
   (f) How does a bat sleep?
   (g) How does a bat hibernate?
   (h) In horror films we often hear of vampires. Do vampires really exist? Where would you find them?
   (i) Name a common species of bat that is found in Ireland.

3 Do a short project on 'The Bat'.

My Record

Draw a mind map on the information that you have about the bat.

My Conclusion

Now that your class has discussed bats, answer this question: Why does the bat hibernate?

The squirrel and the badger

The squirrel and the badger do not hibernate but they often go for a long sleep during cold weather in wintertime. They curl up in their warm, dry nests and go to sleep for a few days. This helps them survive the cold weather.

In third class you probably learned about the badger. Now you can gather together lots of information about the squirrel. Do a short project on the squirrel.

Did you know?

- The biggest frog in the world is called a Goliath frog and it is about the size of a pet cat.
- Some frogs can fly instead of hopping from place to place!

Science and the environment

An echo is very useful in nature and in science.

- The bat finds its way at night by sending out sounds. Echoes return and the bat can then judge what is ahead. This is called **echolocation**. Fishing trawlers use echolocation to find fish in the ocean. They send out sounds that bounce off the fish and the echo tells how far away the fish are.

- Doctors use an **echocardiogram** to examine the human heart. Gel is first applied to a person's skin. Then a device is rubbed over the gel and echoes of the heartbeats appear on a computer screen. The doctor can then clearly see how the person's heart is functioning.

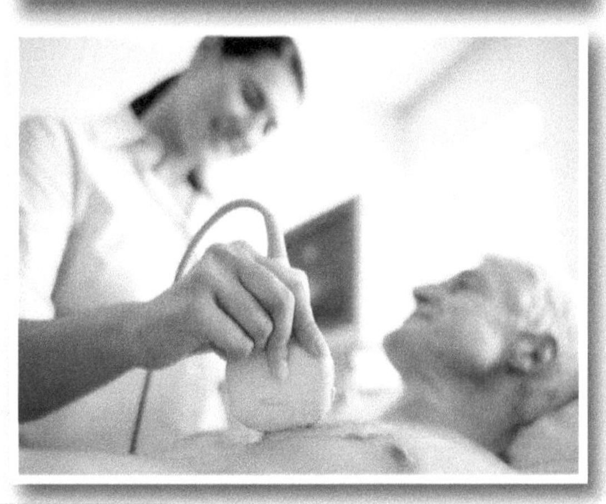

Web Watch!
If you would like to find out more about bats go to this website: http://www.bats4kids.org/

Integration Project

English

1 Write about why you would/would not like to hibernate.
2 Do you think that it would be useful to us human beings if we could hibernate? Why?/Why not?
3 Look up the meaning of the word 'insomnia'.

Gaeilge

Déan pictiúr agus scríobh scéal faoin ghráinneog agus é ag dul a chodladh don gheimhreadh.

Mathematics

If a hedgehog goes to sleep for 3 months (November, December, January), how many days/hours of sleep does it get?

Visual Arts

Make a native American dream catcher to catch your bad dreams:
http://www.enchantedlearning.com/crafts/Dreamcatchers.shtml

Sleepy Heads

Geography

We usually sleep at night and are awake by day. Other cultures may do things differently.
For example, in Spain you might have a siesta – a rest during the hottest part of the day. Find information on countries that have this lifestyle and present your findings to the class.

SPHE

1 List reasons why you think that sleep is important.
2 Did you know that there are four stages of sleep? Find out more about this on:
http://kidshealth.org/kid/stay_healthy/body/not_tired.html

Music

Listen to Brahms' Lullaby. Discuss why you think this music might/might not be suitable for helping to put a baby to sleep.

Science

1 Make a list of all the animals you know that hibernate.
2 Did you ever wonder where insects go during wintertime? Many adult insects hibernate, e.g. the Ladybird Beetle.
Find out more on
http://www.si.edu/encyclopedia_SI/nmnh/buginfo/winter.htm

Why do we eat food?

We eat food because it is nice.

We eat food because we get hungry.

We eat food to help us grow.

We eat food because it gives us energy.

The BIG Question

Identify the main question you need to answer.
Record it now on your Investigation Record Sheet.

My Prediction

Complete this sentence on your Investigation Record Sheet:

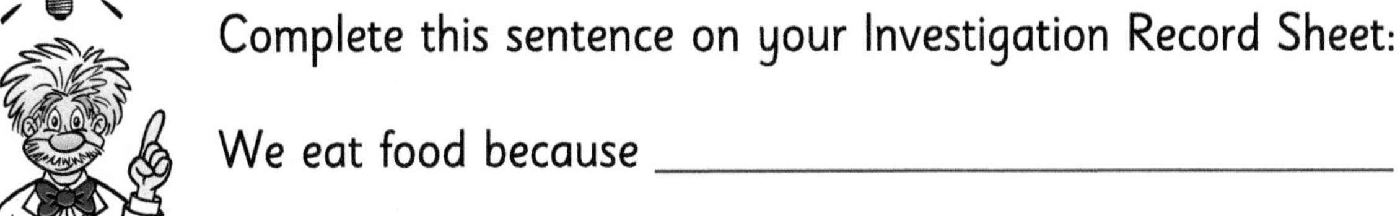

We eat food because ______________________________

____

Let's Investigate!

1 We eat food for many reasons:
(a) Food gives us energy.
(b) Food helps us to grow.
(c) Food helps the body to repair any worn-out parts.
(d) Food helps to protect us against disease.

2 Our bodies need carbohydrates, fats, proteins, vitamins and minerals. These are called nutrients.
  (a) Carbohydrates and fats give us energy.
  (b) Proteins help us to grow. Proteins also help to repair any worn-out parts of the body.
  (c) Vitamins and minerals keep our bones strong, give us healthy teeth and skin and help to keep us healthy.

3 We get different nutrients from different foods:
  (a) Fruit, milk, bread, potatoes, pasta, vegetables and cereals give us **carbohydrates**.
  (b) Cheese, butter, meat and oils give us **fat**.
  (c) Lean meat, nuts, fish and eggs give us **protein**.
  (d) We get **vitamins** from milk, cheese, yoghurt, vegetables and fruit.
  (e) We get **minerals** from liver, eggs, spinach, milk, cheese, fish, fruit, vegetables and drinking water.

4 When we eat the correct amount of all the different nutrients that we require, we are eating a **balanced diet**.

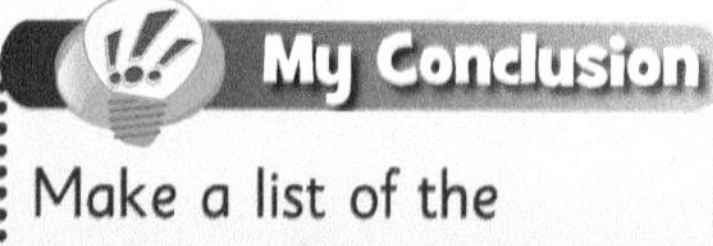

My Conclusion

Make a list of the nutrients that we require. Which foods will give us those nutrients?

Finding Out More

1 Look carefully at the food pyramid on page 25. Look at layer 1. What do you notice about most of the foods on it?

2 Look at layer 2. What do you notice about most of the foods on it?

3 Look at layer 3. What do you notice about most of the foods on it?

4 Look at layer 4. What do you notice about most of the foods on it?

5 Layer 5 contains foods that we should eat only now and again. What do you notice about the foods on it?

In order to eat a balanced diet, we should try to eat from each layer as follows:

Layer 1: at least six servings each day
Layer 2: at least five servings each day
Layer 3: three servings each day
Layer 4: two servings each day
Layer 5: only a little of this food

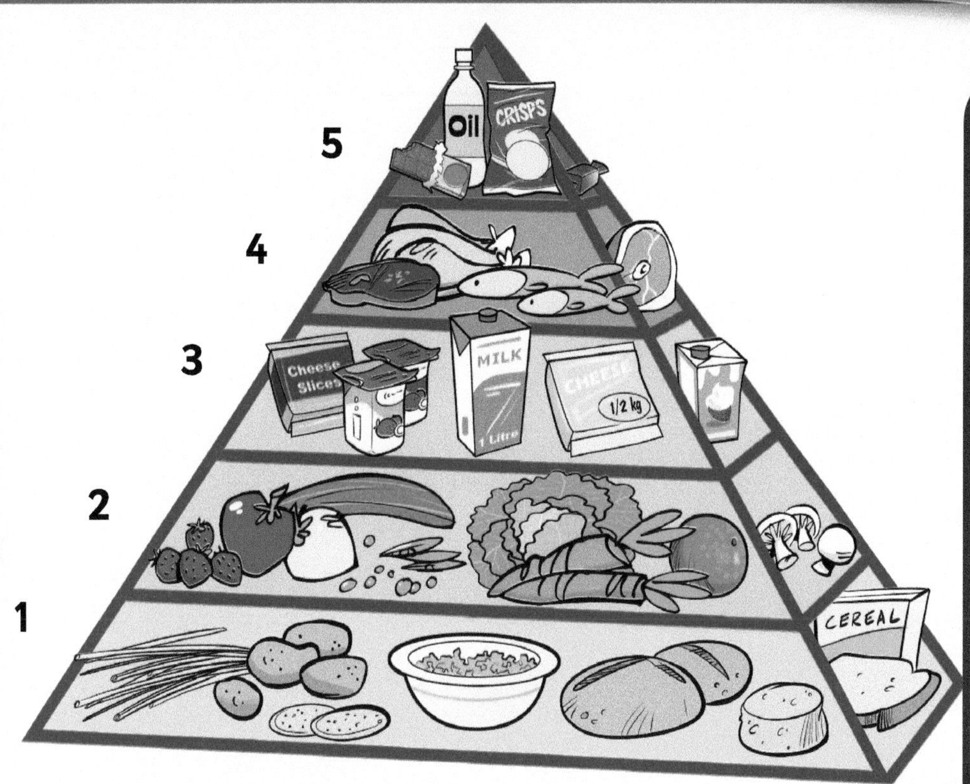

Detective Time

1 Make a list of the food you ate for one day:
(a) Keep a record of all the food you ate for one day.
(b) Examine the list and count how many servings you had from each layer of the food pyramid.
(c) Did you eat a balanced diet for that day?
(d) What changes do you need to make in order to improve your diet?

2 In your copy, write a healthy menu for a day:
(a) Keep in mind the different nutrients that you need for a balanced diet.
- What is on your menu for breakfast?
- What will you include for lunch?
- Provide a snack for when you come home from school.
- What will you have for dinner?

(b) Now look at the foods that you have included in the menu. Is it a balanced diet?
(c) Do you need to add any extras to it? Explain your answer.

Do I eat a balanced diet?
Discuss with your class how each of you could find out if you eat a balanced diet.
Do the investigation.

My Record
Draw diagrams of a food pyramid.

My Conclusion
Do I eat a balanced diet?
Now that your class has discussed diet, you can record your conclusions about your own diet.

My Menu

Breakfast: ____________________
____
Lunch: ____________________
____
Snack: ____________________
____
Dinner: ____________________
____

My Record

Make up a healthy menu for a day. Present it like a menu that you would find in a restaurant.

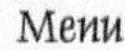

Menu

Starters

Vegetable Soup
Garlic Bread
Seafood Cocktail

Main Course

Chicken
Lamb
Trout

All main courses are served with potatoes and fresh vegetables

Dessert

Ice-cream
Apple Tart and cream
Fruit Salad

My Results

It is now time to record what you ate for the day.

My Conclusion

Now that your class has discussed healthy eating, you can record your conclusions about your own diet.

Very Interesting

Science and the environment

- Carbohydrates give us energy. People who play a lot of sports or take a lot of exercise need to eat extra carbohydrates so that they will have enough energy. Many athletes use **carbohydrate overloading** before big sporting events. They eat a lot of extra carbohydrates so that they will have extra energy when their big day arrives!

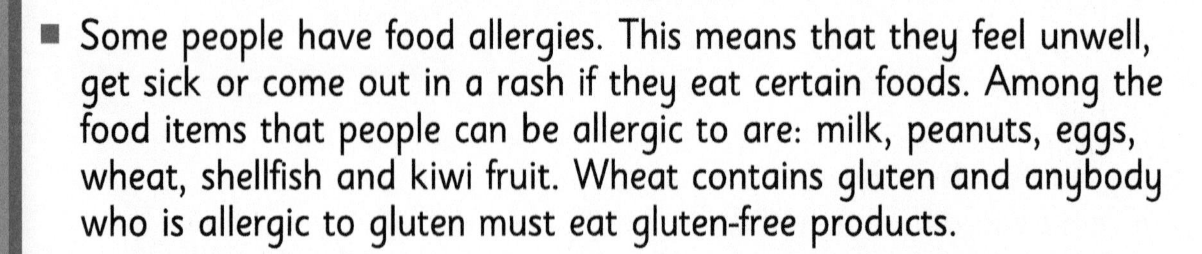

- Some people have food allergies. This means that they feel unwell, get sick or come out in a rash if they eat certain foods. Among the food items that people can be allergic to are: milk, peanuts, eggs, wheat, shellfish and kiwi fruit. Wheat contains gluten and anybody who is allergic to gluten must eat gluten-free products.

- Some people are vegetarians. This means that they do not eat meat. If they do not eat meat, they must make sure that they get enough protein in their diet to remain healthy. Examples of food (other than meat) that contain protein are: soya beans, broad beans, eggs and lentils.

Web Watch!
If you would like to find out more about healthy eating and the food pyramid, go to this website:
http://kidshealth.org/kid/stay_healthy/food/pyramid.html

Integration Project

Food for Life

English

1. Write instructions on how to make a delicious sandwich.
2. Write instructions on how to make a fruit salad.

Gaeilge

1. Scríobh amach na treoracha chun ceapaire folláin a dhéanamh.
2. Scríobh dialann de gach rud a itheann tú i rith lá amháin agus déan comparáid le do chara.

Mathematics

1. If a 50g food product contains 10g protein, 25g carbohydrate and 5g fat, work out the fraction of these in the food product.
2. Find out why **e** is printed after the weight of a food product.

Music

Compose a rap about healthy eating.

History

Learn more about ancient and medieval food on:
http://www.historyforkids.org/learn/food/

SPHE

1. Check three examples of food packaging and read what they say about nutrition.
2. Choose a television advertisement for a food product. Explain why you like/dislike it and why it encourages or discourages you from buying and trying out that product.
3. Explain 'shelf life'. Make a list of foods that have a short shelf life. Then make a list of foods that have a long shelf life.

Science

1. Make a list of foods that you eat and sort them into these categories: High protein, High fat, High carbohydrate.
2. Make lists of foods according to how they are stored/preserved.

Geography

1. Choose a country and research the main types of food eaten there.
2. Examine three food packages and see if they are recyclable or not.
3. Investigate the diet of the first farmers in Ireland.

6 Hot and Cold

Which is the warmest place in the classroom?

I predict that beside the radiator will be the warmest place.

The warmest place will be in the centre of the classroom.

The warmest place will be inside the window.

The warmest place will be on the south-facing side of the room.

Identify the main question you need to answer.
Record it now on your Investigation Record Sheet.

My Question
My Prediction
Let's Investigate!
My Record
My Results
My Conclusion

My Prediction

Complete this sentence on your Investigation Record Sheet:

I think that the warmest place in the classroom is ________

____

Let's Investigate!

You will need:
- Thermometers

1 Find out which is the warmest place in the classroom.

In your group, discuss and make decisions:

(a) Discuss the locations in the classroom where you would like to take the temperature each day for a week.

(b) Make a list of these locations. Do not have more than six locations.

(c) How are you going to measure the temperature?

(d) How many times will you check the temperature each day?

(e) Will you have to check the temperature at the same time each day? Explain your answer.

(f) How will you record your results every day?

(g) At the end of the week you will identify the warmest place in the classroom. It could be interesting to identify the coldest place too. How will you do this?

(h) Record and display your final results. How will you do this?

2 Now it is time to begin. You will need to get a few thermometers. If you don't have one for each location, you can move one thermometer from place to place. If you are using one thermometer, how long will you leave it in each place before recording the temperature?

3 Label each location so that your group will check exactly the same places each day.

4 Begin by taking the temperature on the first day. You could use a table like this one to keep an accurate record:

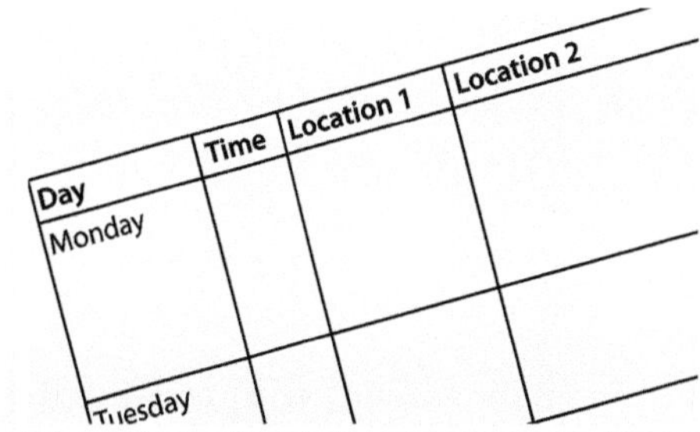

Day	Time	Location 1	Location 2
Monday			
Tuesday			

My Record

On your Investigation Record Sheet, draw diagrams of your investigations.

My Results

It is now time to record your results. Use a table like the one above.

My Conclusion

Now that your class has discussed the investigations, you can record the conclusions.

Finding Out More

You will need:
- Thermometers

This time we are going to do the same investigation but in a different place.

- Talk with your group and decide on a location, other than your classroom, where you would like to find out about the temperature.
- Plan this investigation in exactly the same way as before.
- Make sure that the investigation is carried out accurately, so that you can trust your results.
- Use the same kind of table to record your results.

My Record

On your Investigation Record sheet, draw diagrams of your investigations.

My Results

It is now time to record your results. Use a table like the one you used in the first investigation.

My Conclusion

Now that your class has discussed the investigations, you can record the conclusions.

Detective Time

Discuss these questions within your group. You will have to do some research to answer them.

1 How are our buildings heated?

2 (a) What are fossil fuels?
(b) What fossil fuels are used to heat our homes and schools?

3 (a) What are renewable sources of energy?
(b) What renewable sources of energy are used to heat homes and schools?

4 Which source of energy – fossil fuels or renewable energy – should we try to use whenever possible? Why?

5 What renewable sources of energy are used in Ireland to make electricity?

6 Choose one renewable source of energy and, together with the rest of your group, find out all you can about it. Create a Powerpoint presentation and then show it to the whole class.

My Record

On your Investigation Record Sheet, draw diagrams showing how we use one renewable source of energy to make electricity.

My Results

It is now time to record your results.

My Conclusion

Now that your class has discussed the investigations, you can list reasons why we should use renewable resources of energy to produce electricity.

Science and the environment

Scientists have discovered many ways to use the sun's energy:

- They have invented solar panels. There are two types of solar panel.
  - One type changes light from the sun into electricity. These are often used to power signs on the roadside.
  - The other type uses the sun's energy to heat water. These are often visible on the roofs of houses and other buildings.
- Scientists have invented calculators that are powered by the sun. Which type of solar panel do these calculators use?

Web Watch!
If you would like to find out more about how our atmosphere heats and cools go to this website:
http://www.kidsgeo.com/geography-for-kids/0054-heating-cooling-atmosphere.php

Integration Project

English

Write a story about global warming from the point of view of the polar bear.

Gaeilge

Le cabhair léarscáil na hÉireann, cuir faisnéis na haimsire i láthair don rang.

Mathematics

From the results of your science lesson, create a data chart of the hot spots in your school.

Hot and Cold

Visual Arts

Draw/Paint a picture of an activity or scene. Use 'warm' colours or 'cool' colours to create the atmosphere.

Geography

1 List five ways of saving heat energy. List five ways in which heat energy can be wasted.
2 Think of as many uses as you can for thermometers in our daily lives.
3 Check out the temperature in different parts of the world today:
http://www.mathcats.com/explore/weather.html

SPHE

People are referred to as 'hot headed' if they get into a temper easily.
How would you calm someone down if he/she were having a temper tantrum?

History

Did you know that the Romans invented a form of central heating?
Find out about this and more on
http://www.bbc.co.uk/schools/romans/tech.shtml
and
http://www.brims.co.uk/romans/towns.html

Science

We can get heat energy from the sun. Name some other natural ways of getting heat energy.

What material will I use?

Use your Record Sheet to investigate!
My Question
My Prediction
Let's Investigate!
My Record
My Results
My Conclusion

The BIG Question

Identify the main question you need to answer. Record it now on your Investigation Record Sheet.

My Prediction

Complete this sentence on your Investigation Record Sheet:

I think that the best material to use ____________________

____

Let's Investigate!

You will need:

- A basin
- Water
- A selection of materials
- Graduated cylinders

1. Select materials that you can investigate to see if they will absorb water.
   (a) Talk to your group about the materials that you will choose.
   (b) Make a list and discuss the list with your teacher.

to absorb: to soak up

2. Once everyone is satisfied with the choices, gather the materials.
3. Decide how you will test the different materials. It will be important that the testing is fair.
4. How will you know if the material has absorbed water?
5. Once you have decided how to proceed, get some water and test each material.
6. You could record your results in a table like this one:

Material	Absorbed water	Didn't absorb wate
Tissue paper		
Lollipop stick		
Plastic		
Cloth		
Paper		
Stone		

My Record

On your Investigation Record Sheet, draw diagrams of your investigations.

My Results

It is now time to record your results. Use a table like the one above.

My Conclusion

Now that your class has discussed the investigations, you can record the conclusions on your Investigation Record Sheet.

Finding Out More

You will need:
- Three jars
- Food colouring
- Plastic, e.g. bag
- Paper
- Tissue paper
- A pair of scissors
- A ruler
- A marker

In this investigation we are going to use a different way to find out which materials are the best for absorbing water.

1. Get three jars. **Get some plastic, some paper and some tissue paper.**
   (a) Put a little coloured water into the bottom of each jar.
   (b) Mark the level of water in each jar with a marker.

2. Cut three strips of each of the materials: plastic, paper and tissue paper, so that they are exactly the same. Each strip should measure 1cm by 20cm.

3. Take the strip of plastic and hang it over the edge of one jar.
   Make sure that the bottom of the plastic is touching the bottom of the jar, as in the diagram.
   Now do the same for the paper and tissue paper.

4. Observe each material for five minutes. What do you notice?

5. Look at the level of water in each jar. What do you notice?

6. Which material absorbed the most water?
   Which material absorbed the least?
   How do these results compare with the results of the first investigation?

7. You can investigate other materials if you wish.

My Record

On your Investigation Record Sheet, draw diagrams of your investigations.

My Results

It is now time to record your results.

My Conclusion

Now that your class has discussed the investigations, you can record the conclusions on your Investigation Record Sheet.

Detective Time

You will need:
- Food colouring
- Stick of celery
- Tissue paper
- A magnifying glass

We know that plants need water to survive. Now let's find out:

How does water get into a plant and how does it travel inside a plant?

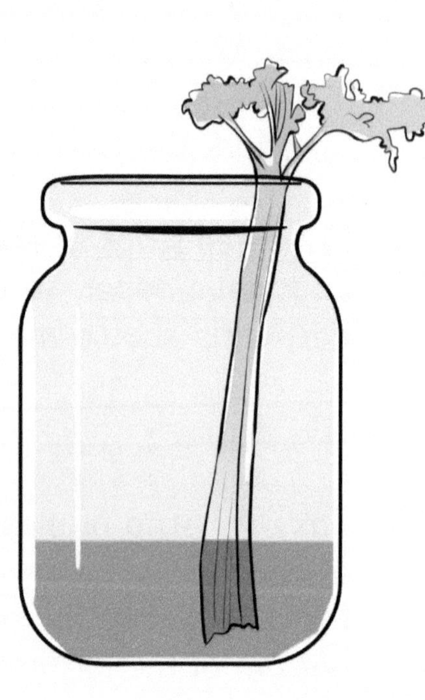

1. Get a stick of celery and wash it so that it is clean. Do not remove the leaves.
2. Get a jar of water and add a little colour to it. Put the stick of celery into the jar as in the diagram.
3. Put the jar on a shelf away from direct sunlight. Leave it for two days.
4. Regularly observe what is happening. What do you notice?
5. Leave it for a little longer. Is there any change?
6. Take out the celery stick and dry it with some tissue paper.
   - (a) Ask your teacher to cut through the stem.
   - (b) Look carefully at the site where the stick is cut. What do you notice?
   - (c) What does this tell you about how water travels up to the leaves?

Compare your celery sticks with these freshly cut sticks.

My Record

On your Investigation Record Sheet, draw diagrams of your investigations.

My Results

It is now time to record your results.

Remember!
- Clear diagrams
- Labels
- Colour

My Conclusion

Now that your class has discussed the investigations, you can record the conclusions on your Investigation Record Sheet.

Very Interesting

Science and the environment

Scientists have developed a useful material – **plastic**.

flexible: bends easily without breaking

rigid: does not bend

- Plastic is a manufactured product. It is made from elements found in oil and natural gas. Different types of plastic are used for different purposes. Some plastic is strong and sturdy, more is light and **flexible**. We use plastic every day. It is used a lot in packaging and in containers.

- It is important that we recycle as much plastic as possible. Recycled plastic is used to make new products. Different types of plastic are recycled to make different products:

  - Plastic that is used in strong bottles, containers and pipes is strong, **rigid** and tough. When recycled, it is made into pipes, containers for oil, containers for detergents and buckets.
  - Plastic used in bags, lids, drinks bottles and toys is flexible, tough and somewhat **transparent**. When recycled, it is used to make plastic pallets and refuse sacks.
  - Plastic sheeting is used to cover many products and is used on farms to cover silage. When it is recycled, it is made into garden furniture.

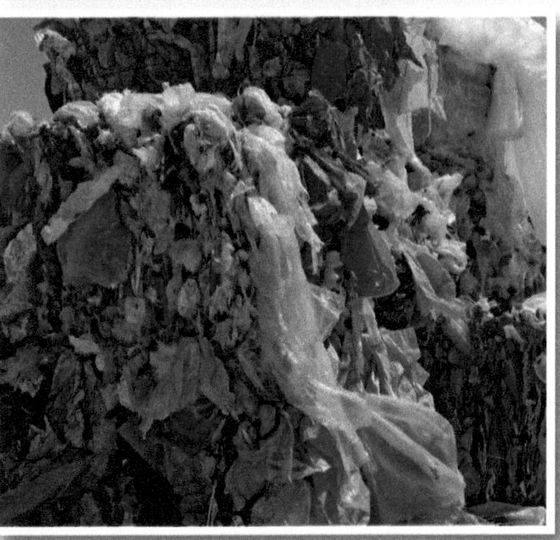

- Recycling plastic uses far less energy than making new plastic, so encourage your friends and family to recycle plastic regularly.

transparent: easy to see through

Web Watch!
If you would like to find out more about the invention of waterproof clothing by Charles Macintosh, go to this website:
http://www.rampantscotland.com/inventors/inventions_waterproof.htm

Integration Project

Soaking Up

English

Write a haiku poem on 'Water'.

Gaeilge

1 Tomhais: Céard atá dubh nuair atá sé glan agus bán nuair atá sé salach?
2 Cad iad na héadaí a chaitheann tú nuair a bhíonn an aimsir fliuch?

Mathematics

1 If one sheet of kitchen paper soaked up 5ml of water, how many sheets would you need to soak up a spillage of 50ml of water?
2 How many sheets would be needed to soak up 105ml of water?

Visual Arts

Design your own pair of funky Wellington boots.

History

The first Duke of Wellington, Arthur Wellesley, made wellington boots popular.

Investigate the history of the Wellington boot.

SPHE

List the things that your body needs to absorb in order to stay healthy.

Physical Education

Welly throwing is a sporting event which originated in Britain. Find out more about it.

Geography

Oil spills pose a major threat to sea life. Learn more about this on:
http://library.thinkquest.org/CR0215471/oil_spills.htm

Science

We use absorption materials every day. List as many situations as you can where you would use them around the home.

Will the switch turn off that light?

Identify the main question you need to answer.
Record it now on your Investigation Record Sheet.

My Prediction

Complete this sentence on your Investigation Record Sheet:

I think that the best way to stop the bulb lighting is to

____

Let's Investigate!

You will need:
- A battery
- A bulb
- A battery holder
- A bulb holder
- Wire

1. Set up a simple circuit with a battery, a battery holder, a bulb and a bulb holder. Get the bulb to light. Be very careful when working with a bulb and wire.

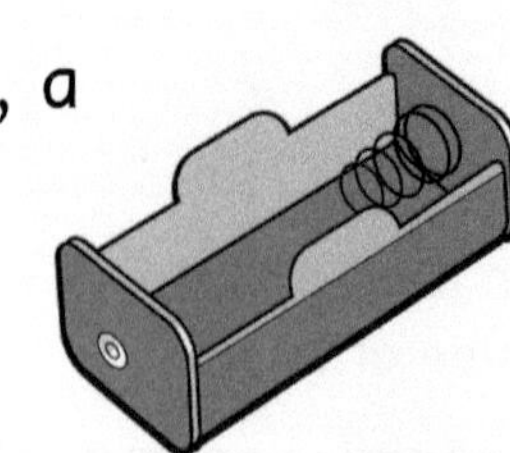

2. Discuss with your group how you might be able to stop the bulb lighting. Make a list of the suggestions.

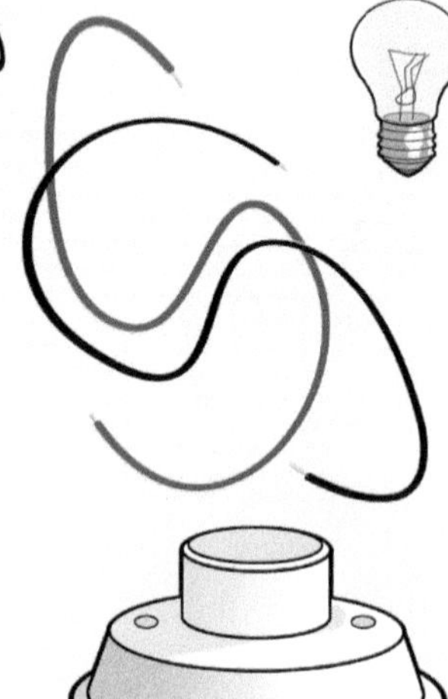

3. Try out all of the suggestions and identify the ones that work.

4. You could use a table like this one to record what your group did:

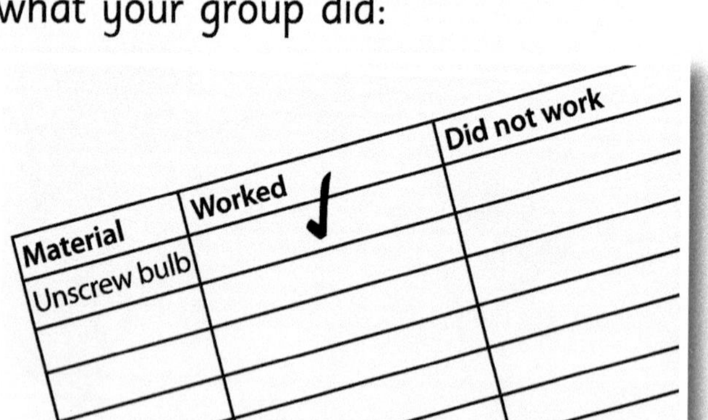

Material	Worked	Did not work
Unscrew bulb	✓	

My Record

On your Investigation Record Sheet, draw diagrams of your investigation.

My Results

It is now time to record your results. Use a table like the one above.

Remember!
- Clear diagrams
- Labels
- Colour

My Conclusion

Now that your class has discussed the investigations, you can record the conclusions on your Investigation Record Sheet.

Finding Out More

<table><tr><td>

You will need:
- A battery
- A bulb
- A battery holder
- A bulb holder
- Wire
- A bell switch
- A push switch

</td></tr></table>

1 Get a **bell switch** from your teacher.
Remake the circuit and put the switch into it.

2 Find out if you can use the switch.
Does it work?
Explain what you need to do to turn on and off the bulb.

3 Most manufactured switches are sealed and you cannot see inside them.
Look carefully at the bell switch that you are using.
How do you think it works?

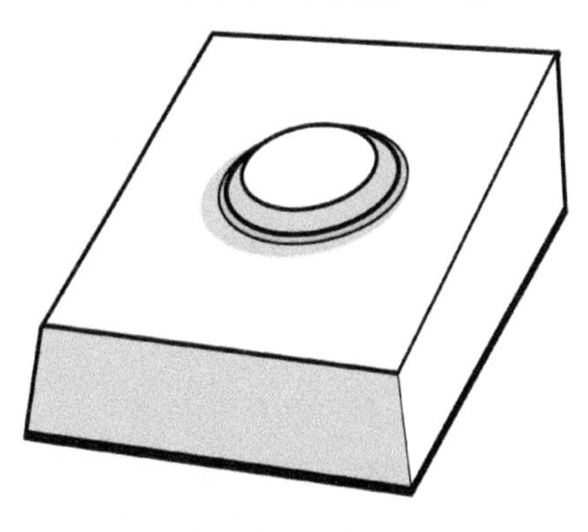

Bell switch

4 Get a **push switch**. Put it into the circuit instead of the bell switch.

5 Find out if you can use this switch. Does it work?
What do you need to do now to turn on and off the switch?

6 This switch is sealed and you cannot see inside it.
How do you think it works?

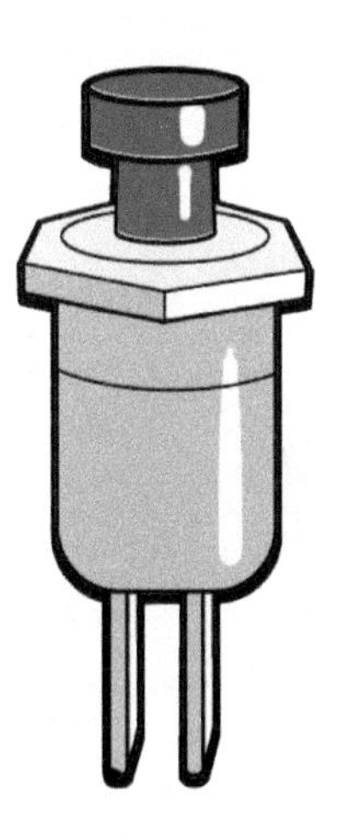

Push switch

My Record

On your Investigation Record Sheet, draw diagrams of your investigations.

My Results

It is now time to record your results.

My Conclusion

Now that your class has discussed the investigations, you can record the conclusions on your Investigation Record Sheet.

Detective Time

1. In your group, discuss how you might go about making your own switch.

2. You will need to get some suitable materials such as cardboard, two split pins and a paper clip. Look carefully at the diagrams and see if you and your group can make a simple switch.

You will need:
- Cardboard
- Two split pins
- A paper clip
- A battery
- A battery holder
- Wire
- A bulb
- A bulb holder
- Scissors

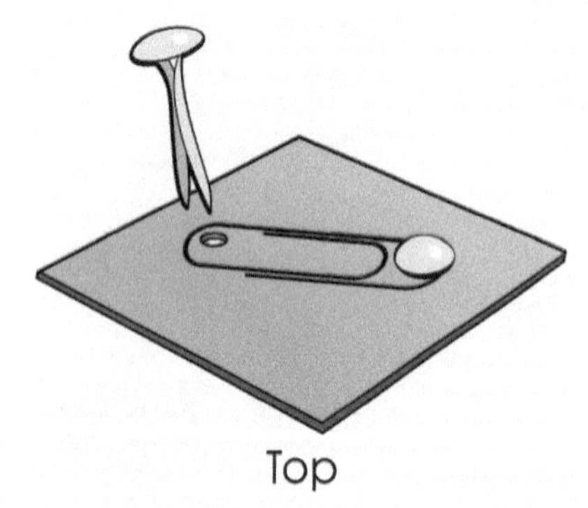

Top

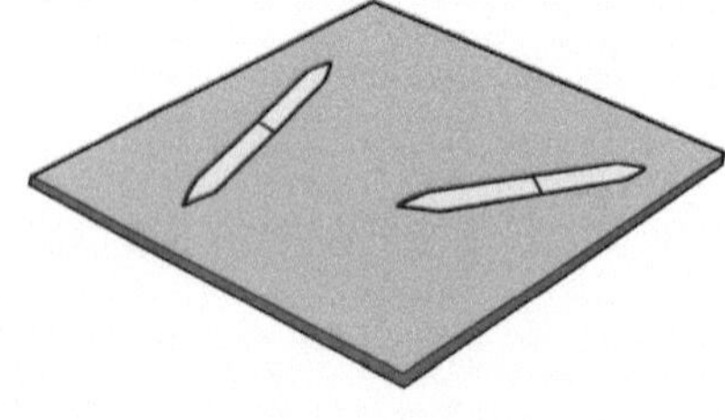

Underside

3. Once you have the switch made, see if you can put it into a circuit.

4. Turn on the light and then turn it off.

5. If your switch isn't working well, you may need to change it slightly.

6. Talk with your group and see if you could make any changes to the switch that would improve it in any way.

My Record

On your Investigation Record Sheet, draw diagrams of your investigation.

My Results

It is now time to record how you made a switch.

My Conclusion

Now that your class has discussed the investigations, you can record how a bulb is turned on or off by a switch.

Make and Do

Work with your group to design and make a lighthouse.

1. Begin by drawing a design of the lighthouse.

2. Make a list of all the materials that you will need.

3. Draw a diagram of the wiring needed to make the bulb work on top of the lighthouse.
   Where will you store the battery?
   Will you use a switch? Which type?

4. Put your finished lighthouse on display in the classroom.

Science and the environment

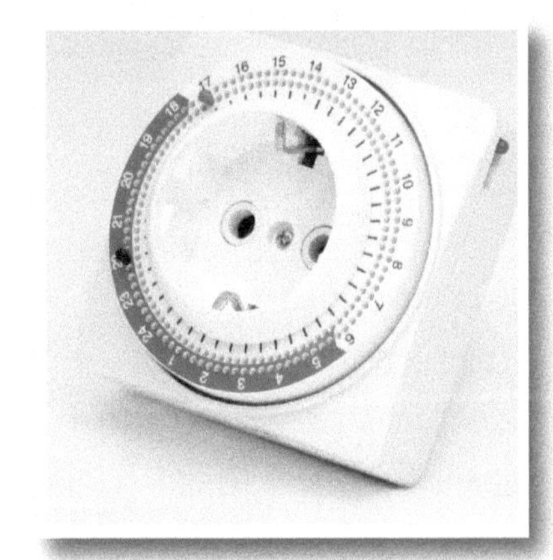

- Scientists have invented different types of switches, which we use to make our lives easier. **Timer switches** are fitted to some machines so that the machines will work only when needed. Can you think of any such machines or appliances?

- Sometimes it is nice to be able to reduce the light in a room, so scientists have invented **dimmer switches**. These reduce the flow of electricity to the bulb, so that it shines more dimly.

Web Watch!
If you would like to find out more about circuits, go to this website:
http://www.hantsfire.gov.uk/circuits

If you would like to see pictures of lighthouses in Ireland, go to this website:
http://indigo.ie/~eaglejr/lhhook.html

Integration Project

English

Switching: choose a topic and start off writing your story on a page.
At a signal from your teacher, switch pages with the pupil to your right.
Read what is written on the page given to you and, until the next signal, continue the story in a way that makes sense. Repeat until the last signal.
Read out the stories to the class. Can you recognise your own story starter?

SPHE

Compose an advertisement convincing someone to switch from fossil fuels to renewable energy.

Gaeilge

Scríobh síos cén sórt aimsire a bhíonn ann i rith na séasúir –
an t-Earrach, an Samhradh, an Fómhar agus an Geimhreadh.

Switch it on!

Science

1. Investigate different types of switches.
2. Look around you and make a list of anything that needs a switch in order to function.

Mathematics

1. A timer switch was programmed to activate a lamp at the following times: 3am, 5am, 6.45am, 4.30pm and 10.15pm. What are these times on a 24-hour clock?
2. The lamp stayed on for 20 minutes each time. For how long, in total, was the light on?

History

When electricity became available in Ireland, people switched over to using it for light. How did people light their houses before they used electricity?

Geography

List ways in which a timer switch could help us conserve energy and make our environment safer.

Will the bulb light if I place an extra battery in the circuit?

Identify the main question you need to answer.
Record it now on your Investigation Record Sheet.

Use your Record Sheet to investigate!
My Question
My Prediction
Let's Investigate!
My Record
My Results
My Conclusion

Complete this sentence on your Investigation Record Sheet:

I think that the voltage of the bulb must be

voltage: the number of volts of electricity the bulb can hold

You will need:

- Wire
- Two bulbs (1.5v and 3v)
- Two batteries (1.5v)
- Two battery holders (a single and a double)
- Switches
- A bulb holder
- A screwdriver

We will set up a simple circuit with a battery, a battery holder, a bulb and a bulb holder. Be very careful when working with a bulb and wire.
We will put a push switch into the circuit.

1. Look at the battery. What voltage is it?
2. Now look at the bulb, especially at the metal part. What voltage is the bulb?
3. Put a 1.5v battery into the holder. Put a 1.5v bulb into the bulb holder.
   Turn on the light. What do you notice?
4. Replace the bulb with a 3v bulb.
   Turn on the light. What do you notice?
5. Replace the battery holder with one holding two batteries, so that you have 3 volts.
   Again, turn on the light. What do you notice?
6. Replace the bulb with a 1.5v bulb.
   Turn on the light. What do you notice?

My Record

On your Investigation Record Sheet, draw diagrams of your investigations.

My Results

It is now time to record your results.

Remember!
- Clear diagrams
- Labels
- Colour

My Conclusion

Now that your class has discussed the investigation, you can record the conclusion on your Investigation Record Sheet.

Finding Out More

You will need:
- A double battery holder
- A connection with wires
- Two batteries
- A bulb
- A bulb holder

1. Examine a double battery holder. Insert the batteries in the correct way.
2. Examine the way in which the metal strips in the holder are arranged. Follow the route the current will travel from battery to battery and then through the wires. Why are the batteries placed in opposite directions?
3. Draw a diagram of the circuit that the electricity follows within the battery holder.
4. The wires coming from the battery are two different colours. Why is this?

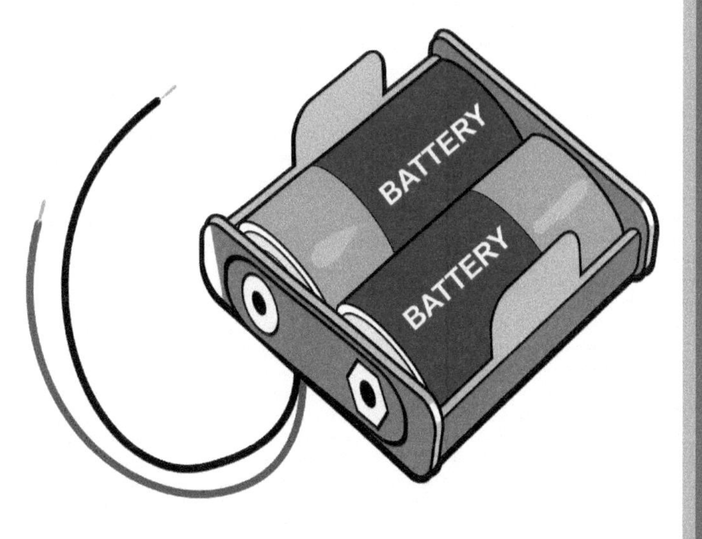

My Record

On your Investigation Record Sheet, draw a diagram of a double battery holder with no batteries in it.

My Results

It is now time to answer the questions above.

My Conclusion

Now that your class has discussed the investigation, you can record how a double battery holder works.

Detective Time

You will need:
- Wire
- A buzzer
- A torch
- A battery
- A battery holder

1 How does a buzzer work?
Get a buzzer and see if you can set up a circuit in which it will work.

2 How does a torch work?
- Get a torch. Is it working?
- Open it and carefully examine how the batteries are placed inside.
- Take out the batteries.
- Look at the bulb and see how it is fitted into the torch.
- Look at the switch that is used. Can you see how it works?

My Record

On your Investigation Record Sheet, draw diagrams of your investigations.

My Results

It is now time to record your results.

Remember!
- Clear diagrams
- Labels
- Colour

My Conclusion

Now that your class has discussed the investigations, you can record the conclusions on your Investigation Record Sheet.

Very Interesting

Science and the environment

- Scientists have invented the **fuse**. This protects a circuit. If too much electrical current flows in a circuit, the metal wire in the fuse melts and breaks the circuit. This is a safety feature.
- Can you think of any safety risks in the household to do with electricity, switches, bulbs and batteries? What safety features could be used?

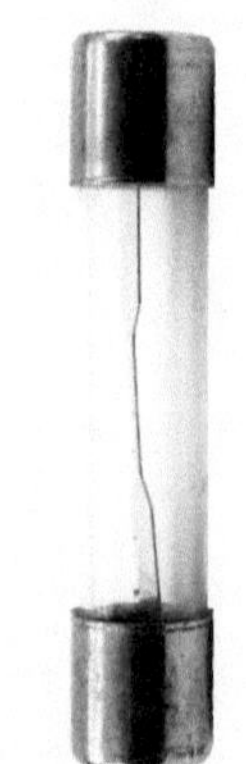

Web Watch!
If you would like to find out more about making electric circuits, go to this website: http://www.zephyrus.co.uk/circuits1.html

Integration Project

English

Brainstorm words to do with electricity and circuits. Make a list and compare with those of your classmates.

Gaeilge

Déan liosta de na rudaí a úsáideann leictreachas i do sheomra ranga.

Mathematics

1 If the length of a building is 30m and the width is 10m, what is the perimeter of the building?
2 Find out the perimeter of your school. Make string 'metre sticks' measuring 10m and 1m. Work in pairs. Estimate first and compare your findings with your classmates.

Visual Arts

1 Design a maze and challenge a friend to get to the treasure.
2 Take a line for a walk on a page, completing a circuit with lots of loops. Use bright colours to shade in the loops.

Batteries and Bulbs

History

Many of the games that we have nowadays involve batteries and electricity.
Ask your parents and grandparents about games and pastimes that they enjoyed and that did not need this technology.

SPHE

Make a circle with your group. Pass a hand-shake around the circle. Pass a smile around the circle.

Science

1 Make a list of items that use electricity.
2 Make a list of items that use a battery.
3 Make a list of items that can use either electricity or a battery.

Geography

Design a circuit in your area for a bicycle race. Mark in features to help the cyclists identify where they are going. Use arrows to show the correct direction around the circuit.
Show where there is water/fruit available to cyclists on the journey.

10 The Nature of Things

Is it natural or manufactured?

It is natural because it is made from wood.

It is manufactured because it was made in a factory.

It is both natural and manufactured.

It is manufactured because the varnish and materials used to finish it are not natural.

The BIG Question

Identify the main question you need to answer.
Record it now on your Investigation Record Sheet.

My Prediction

Complete this sentence on your Investigation Record Sheet:

I think that ______________________________

Let's Investigate!

You will need:
A selection of materials: a stone, a lollipop stick, a lunch box, a pencil, a pair of scissors and a book

1 Go to your teacher and get a selection of items. Look at them carefully. Discuss with your group how you are going to classify them.

2 Make two groups of the materials.

3 You could use a table like this to record your decisions:

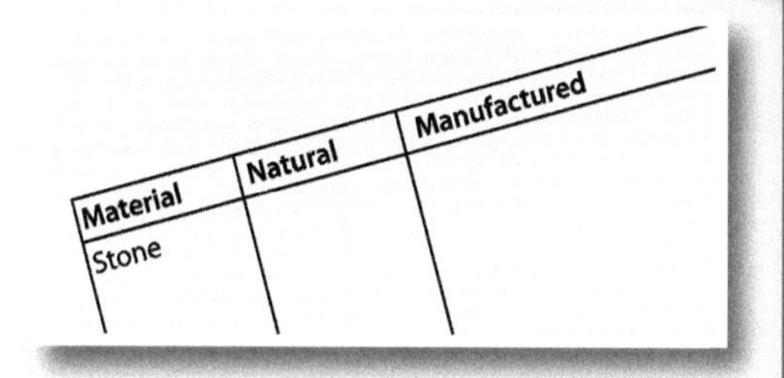

Material	Natural	Manufactured
Stone		

My Record

On your Investigation Record Sheet, draw diagrams of some of the materials you used.

My Results

It is now time to record your results. Use a table like the one above.

My Conclusion

Now that your class has discussed the investigation, you can record the conclusion.

Finding Out More

- In our first investigation we grouped materials as either natural or manufactured.
- This time we are going to look all around the classroom and even around the schoolyard.
- We will make a list of as many objects and materials as possible around the school.
- These items will then be grouped according to what they are made from.

1 Talk to your teacher and organise to look carefully all around the school and find as many different objects as you can. Make a list.

2 When everybody comes back, look at the list and then decide how to group the objects.

3 How are you going to record your results?

My Record

On your Investigation Record Sheet, draw diagrams of some materials made from wood.

My Results

It is now time to record your results.

My Conclusion

Now that your class has discussed the investigation, you can make a list of items made from more than one material.

Detective Time

We use different types of materials to do different jobs for us.

1 Our clothes

(a) In winter, when it is cold, what types of clothes do we wear?
(b) Why do we wear these types of clothes?
(c) In summer, when it is warm, what types of clothes do we wear?
(d) Why do we wear these types of clothes?

2 Our homes

(a) It is important to insulate our homes. What does 'insulate' mean?
(b) What materials do we use to insulate our homes? Make a list.
(c) Are there any other ways in which we insulate our homes?
(d) Why is it important to insulate our homes?

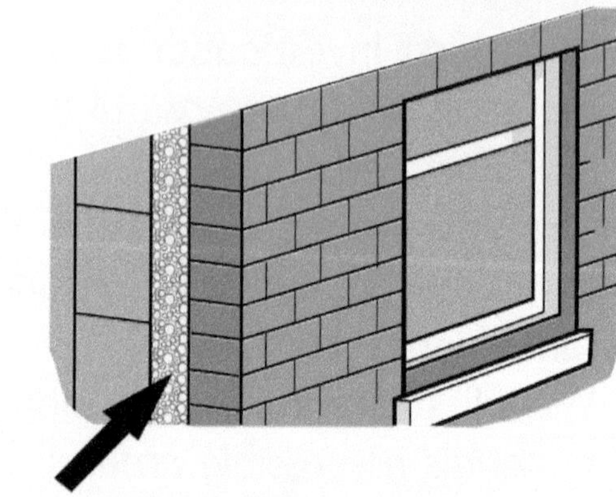

Cavity insulation

3 Our food

(a) It is important to keep our food fresh.
(b) How do we keep our food fresh?
(c) Scientists have done a lot of research to help us keep our food fresh. They have invented a number of ways to preserve food. Talk about it with your group. You may need to go to the internet or an encyclopaedia to research food preservation.
(d) Make a list of ways in which we preserve food.

My Record

On your Investigation Record Sheet, draw diagrams of different types of materials.

My Results

It is now time to record your results.

My Conclusion

Record the conclusion on this question: Why do we use different materials to do different jobs?

Very Interesting

Science and the environment

- Scientists often mix materials to make a new product. This mixing improves a product for the job that it has to do.
  - Iron and carbon are mixed to make steel.
  - Various chemicals are mixed to make a strong, light cloth called nylon.
  - Copper and tin are mixed to make bronze.
  - Sand, lime, soda and other products are mixed to make glass.

- Scientists in New York and London worked together to make nylon. They invented the name for this new material. Find out how they decided on the name.

Nylon rope

Web Watch!
If you would like to find out what ecokids do, go to this website:
http://www.binhaitimes.com/

Integration Project

English

1. Use a Thesaurus to find as many words as you can for the word 'manufacture'.
2. Write out the arguments for or against the motion 'It is right that organic foods should cost more than non-organic foods'. Hold a class debate on this issue.

Gaeilge

Déan liosta de na rudaí gur féidir a úsáid chun anraith glasraí a dhéanamh.

Mathematics

1. If 5 bags of fibreglass could insulate the attic of a house, how many bags would be needed to insulate 12 houses?
2. If 45 sheets of styrofoam could insulate the cavity of a house, how many houses would 495 sheets insulate?

SPHE

1. Examine labels on your clothing and see how many fabrics are used. 100% wool means that the item of clothing is made from wool only.
2. Find out what 'organic' means. Check out the organic section in your local supermarket. List any advantages and disadvantages of eating organic foods: http://www.kidshealth.org/teen/food_fitness/nutrition/organics.html

The Nature of Things

Geography

Many people would argue that climate change is caused by humans. Would you agree with this? Give your reasons.

History

Find out more about the industrial revolution on: http://www.kidinfo.com/American_History/Industrial_Revolution.html

Science

Industry manufactures many products that we use every day. List some of these products used by you and your family.

A Cosy Nest

Why do birds build nests?

They lay eggs in a nest.

They use the nest as their home.

They show off to their friends by building a nice nest.

They store food in their nest.

The BIG Question

Identify the main question you need to answer.
Record it now on your Investigation Record Sheet.

My Prediction

Complete this sentence on your Investigation Record Sheet:

I think that birds build nests because ________________

Let's Investigate!

In Third class you learned about the thrush, the blackbird and the robin.

- What can you remember about each bird?
- Put together a short presentation with some information on each bird.
- Show this to your class.

Now we are going to find out about the type of nest that each one builds.

The thrush

1 The thrush

(a) Look carefully at the thrush's nest. What shape is it?

(b) The nest is usually built in a bush or in a hedge. The female builds the nest very early in spring. It is made from grass, moss, leaves and twigs. When it is shaped and finished, the female lines the inside of it with cow dung. Why does she do this?

(c) What do you think the male does, while the female is building the nest?

2 The blackbird

The blackbird's nest

(a) Look carefully at the blackbird's nest. What shape is it?

(b) The blackbird builds its nest in trees, in bushes and sometimes in old buildings. The female gathers grass, leaves and twigs to make the nest. She uses mud to keep the nest together.

(c) Do the male and female blackbirds look the same? Describe the female blackbird.

Male and female blackbird

3 The robin

(a) Look carefully at the robin's nest. What shape is it?

(b) The nest is usually built on a small tree or in a hedge. It can sometimes be found in a crack in the wall. The female builds the nest, using grass and twigs. It is not unusual to find a robin's nest in an old box or in a piece of refuse, especially if it is in a quiet, sheltered place. Why would the refuse need to be in a quiet, sheltered place?

The robin's nest

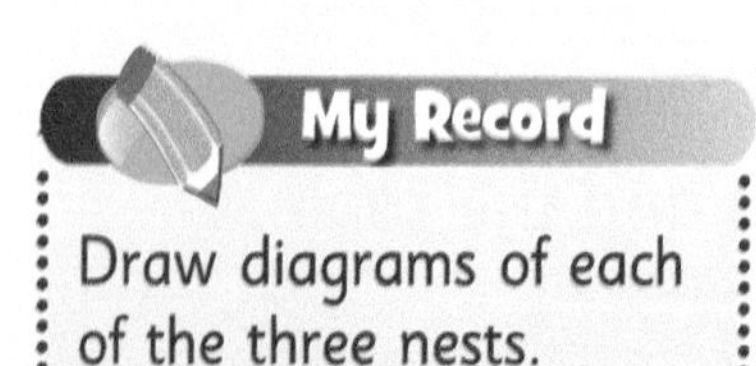
My Record

Draw diagrams of each of the three nests.

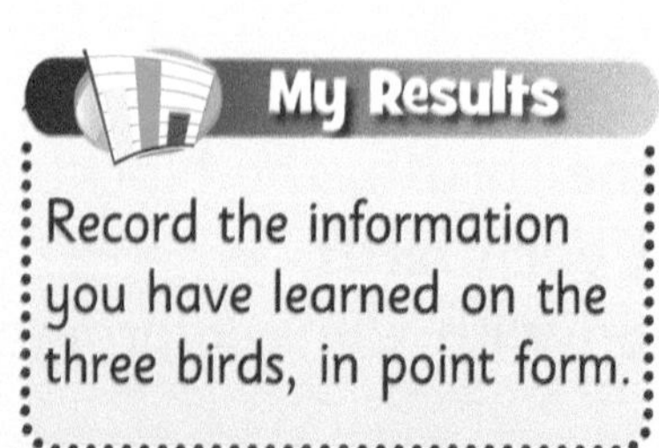
My Results

Record the information you have learned on the three birds, in point form.

Finding Out More

The swallow

- The swallow is a very distinctive bird. Swallows arrive in Ireland in great numbers in April and they are gone again by the end of September. They are often associated with the summer because they arrive at the start of summer and leave when the summer ends.
- Where do swallows spend the rest of the year? Why do they spend our wintertime there?
- Swallows live entirely on insects, which they swallow as they fly.
- Swallows nest inside barns, under the roofs of houses or inside an unused shed or store.
- Look at the nest. What shape is it? The inside of the nest is lined with feathers.
- The male and the female work together in making the nest and in taking care of the young. They can rear up to three families in one season.
- Swallows often return to the same nesting place year after year.

The swallow

A swallow's nest

The house martin

- House martins look like swallows but you will always be able to distinguish them by looking at the tail and the neck.
- Look carefully at the diagrams of the swallow and the house martin. What is the difference between them?

The swallow

The house martin

My Record

Draw diagrams of the swallow and the house martin.

My Results

Record the information you have learned on the swallow, in point form.

Detective Time

The magpie

- The magpie is a large bird that can be found in many habitats. Magpies like to live in woodlands but they can also be found around houses, in small gardens or in housing estates.

- Magpies usually find their food on the ground. They eat beetles, insects, nuts and fruits. In cities they are great scavengers. They will attack dustbins and eat dead animals and birds. In the nesting season, they will eat the eggs and young of other birds.

- Look carefully at the magpie's nest. What shape is it? Where is it usually built?

- Magpies begin to build their nests as early as January. These are made from big sticks and are lined with mud. Magpies often build a roof of sticks over the nest. They lay their eggs in April.

The magpie

The magpie's nest

Magpie hatchlings

My Record

Draw diagrams of the magpie and the magpie's nest.

My Results

Record the information you have learned about the magpie, in point form.

Did you know?

- Magpies have been in Ireland since about 1670. It is thought that a group of about a dozen of them were blown by a strong east wind into Co. Wexford and, over the years, they have spread all over the country.
- Young swallows eat about 400 meals each day, so their parents are kept very busy bringing food to them!

Science and the environment

The corncrake

- We know that some birds are in danger of extinction. Scientists have made us all aware of the work that needs to be done to protect these birds. In Ireland the corncrake is in danger. This bird is protected in certain areas and scientists have told the people who live there how to keep the corncrake safe. Farmers are encouraged to farm in such a way that they will not interfere with the nesting habits of the corncrake.
- Scientists have identified Special Areas of Conservation (SAC) throughout our country and have told us how to protect these areas. Find out more on this website: www.epa.ie/environment/biodiversity

Web Watch!
If you would like to find out more about Irish birds go to: www.askaboutireland.ie/learning-zone Go to the environment section, click on 'feathered friends' and you can find out about lots of birds.

There is also lots of information about Irish birds on: http://homepage.eircom.net/~edrice/birds/index.htm

Integration Project

English

Start off with the word NEST. See how many words you can make by changing just one letter each time.
Take the challenge with a friend and see who can make the most words in a chosen length of time.

Gaeilge

Cén Gaeilge atá ar 'blackbird', 'swallow', 'magpie', 'thrush' agus 'robin'?
Scriobh sliocht gearr ar ceann amháin de na h-éin seo.

Mathematics

1 Can you think of counting rhymes/ songs?

2 If a magpie gathers 5 branches from 4 different trees, how many branches has it gathered?

Music

Find out about the dawn chorus and birdsong:
http://www.rte.ie/radio/dawnchorus/birdsong.html

A Cosy Nest

History

Illustrate the traditional rhyme 'One for sorrow'.

SPHE

Eggs are an important source of protein in our diet. Can you list other foods that provide a good supply of protein?

Geography

Hedge-cutting and tree-cutting along the roads are allowed only at certain times of the year, so as to protect the wildlife living there. Think of other ways in which we can protect a bird's habitat.

Science

Learn by heart:
The cuckoo comes in April,
He sings his song in May,
In June he changes his tune
And in July he flies away.

How will I help the man to hear?

I will help him to hear if I shout louder.

I will try to get him a hearing aid.

I will stand in front of him, so that he will be able to read my lips.

I will move very near to him when I speak.

Identify the main question you need to answer.
Record it now on your Investigation Record Sheet.

Use your Record Sheet to investigate!
My Question
My Prediction
Let's Investigate!
My Record
My Results
My Conclusion

My Prediction

Complete this sentence on your Investigation Record Sheet:

I think that the best way to help the man is to

____

Let's Investigate!

You will need:
- An A4 sheet of paper
- Sticky tape

1. Get a sheet of A4 paper and twist it into a cone shape, as in the diagram.
   Make the cone as wide as possible at one end and just wide enough for your mouth at the other end.
   Use sticky tape to hold it in shape.
   Into which side of the megaphone should you speak, in order to use it? Discuss this with your group.

2. What do you think will happen if you speak into the megaphone? How would you test the megaphone to see if it works?

3. Design a test for inside the classroom.
   Then design a test for outside in the schoolyard.
   Discuss the possibilities with your group.

4. Test the megaphone inside the classroom.
   What did you find?

5. Test the megaphone outside the classroom.
   What did you find?

6. Would the megaphone be of any use to the elderly man?

My Record

On your Investigation Record Sheet, draw diagrams of your investigations.

My Results

It is now time to record your results.

Remember!
- Clear diagrams
- Labels
- Colour

My Conclusion

Now that your class has discussed the investigation, you can record the conclusion on your Investigation Record Sheet.

Finding Out More

You will need:
- An A4 sheet of paper
- Sticky tape

This time we are going to make an ear trumpet.

1 Get a sheet of A4 paper and twist it into a cone shape, as in the diagram.
Make the cone as wide as possible at one end and just wide enough to fit into the outer part of your ear at the other end.
Use sticky tape to hold it in shape.
Can you work out how to use the ear trumpet? Discuss it with your group.

2 Design a test to find out if the ear trumpet would help the elderly man.
Discuss it with your group.
What did you decide?

3 Now do the test.
Did it make it easier to hear?

Remember!
- Clear diagrams
- Labels
- Colour

My Record

On your Investigation Record Sheet, draw diagrams of your investigation.

My Results

It is now time to record your results.

My Conclusion

Now that your class has discussed the investigation, you can record the conclusion.

Unscramble these words:

1 rea ____________	**2** nma ____________
3 rmtuept ____________	**4** noec ____________
5 amgepeonh ____________	**6** spake ____________
7 afed ____________	**8** erhiagn ____________

Detective Time

Discuss within your group what the difference might be between a high-pitched sound and a low-pitched sound.

You will need:
- Wood
- Three drawing pins
- A selection of elastic bands

1. Get a piece of wood and put three drawing pins into it, as in the diagram. You have made the three corners of a triangle. Place the drawing pins in such a way that one side of the triangle will be short, another a little longer, and the third one a lot longer.

2. Choose a few elastic bands of the same length but of different widths. Stretch the narrowest band around the drawing pins, as in the diagram. Make sure that they are tight when in place.

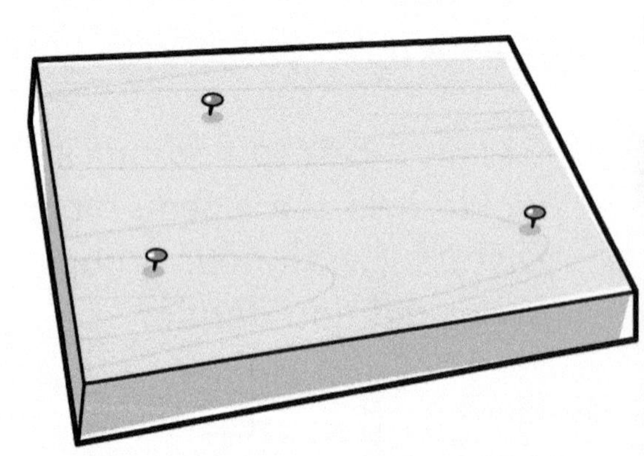

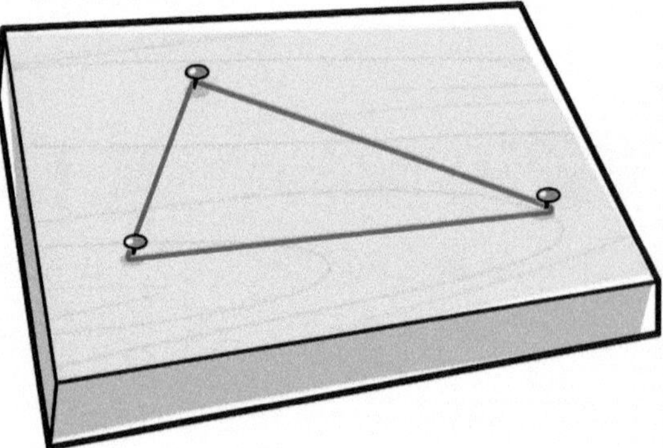

3. Pluck the elastic band on the longest side of the triangle. Then pluck the band on the shortest side. What do you notice about the sounds that come from the long and short bands?

4. Now test the third part of the elastic band. What do you notice?

5. Take off that elastic band and get another one of a different width. Put that band in place. Test the sounds of the different sides. What do you notice this time?

6. Repeat a few times. What do you notice?

My Record

On your Investigation Record Sheet, draw diagrams of your investigations.

My Results

It is now time to record your results.

My Conclusion

Now that your class has discussed the investigations, you can record the conclusion.

Very Interesting

Science and the environment

- Sounds can help keep us safe. Traffic lights at pedestrian crossings have a buzzer. It goes off when it is safe to cross, which is very helpful to blind people. Large vehicles have buzzers, which go off to warn us whenever the vehicle is reversing. Can you think of other sounds that are used as a warning in order to increase our safety?

Web Watch!
Listen to some well-known classical music and learn about famous composers on:
http://www.dsokids.com/listen/composerlist.aspx

English

Complete this story:
'I woke suddenly to the sound of a burglar alarm ...'

Gaeilge

Déan liosta de na torainn a chloiseann tú i rith an lae ar scoil.

Mathematics

If a CD with 10 songs cost €15, how much cheaper would it be to download the songs from *itunes* at 99c per song?

Music

Make a list of your favourite songs.

Loud and Clear

History

Town criers were appointed long ago to deliver important news to the citizens of the country. Find out more on:
http://www.atspromotions.com/towncrier/history.htm

SPHE

Nowadays coping is made easier for people who are deaf or hard of hearing. Can you think of a few examples?

Geography

1 List the sounds within buildings that alert us to danger.
2 List the sounds in the outside environment that alert us to danger.

Science

List all of the sounds you can hear right now. Compare your list with your partner's list.

13 Nature's Ploughman

What is this creature?

I think that it is a small snake.

It looks like an earthworm.

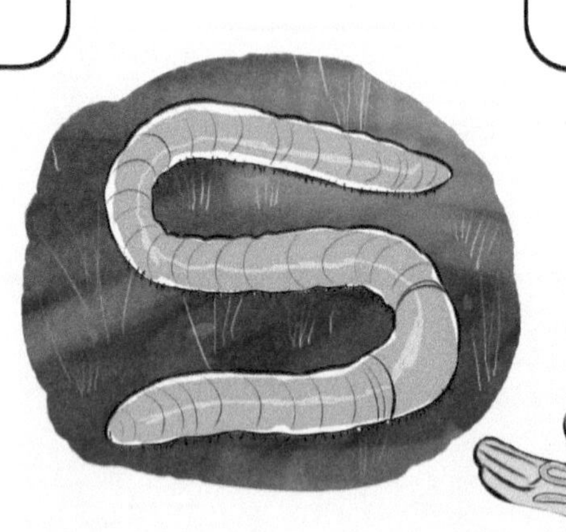

It is a maggot.

It is a type of snail that has no shell.

The BIG Question

Identify the main question you need to answer.
Record it now on your Investigation Record Sheet.

My Prediction

Complete this sentence on your Investigation Record Sheet:

I think that the work of an earthworm is

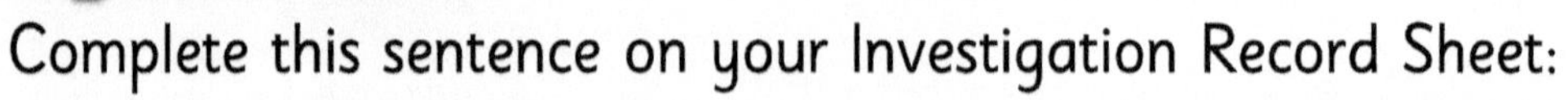

Let's Investigate!

1 An earthworm lives in the soil. It has a long, wet body.
   (a) Describe the body of the earthworm.
   (b) If you look carefully, you will see a section that is thicker than the rest of the body. This is called the saddle.

2 An earthworm has no legs.
(a) Can it move?
(b) How does it move?
(c) If you look carefully, you will see that it has tiny bristles underneath its body. These also help the earthworm to move.

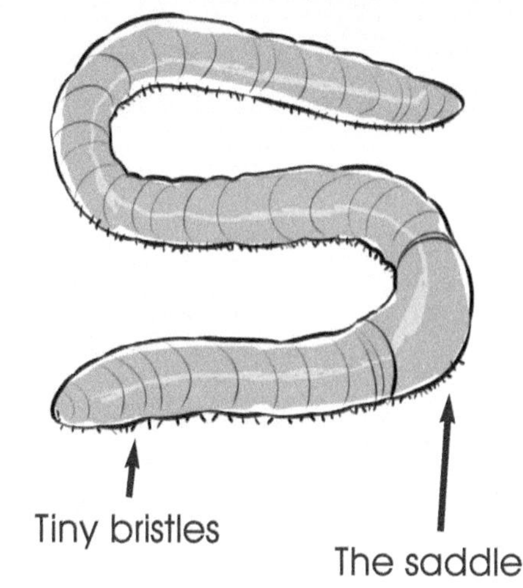

3 An earthworm is blind and deaf. Even so, it can feel the slightest movement in the earth around it. It has a mouth but it has no teeth and no tongue.
(a) What does it eat?
(b) How does it eat its food?
(c) There are thousands of earthworms in every garden. They are always making tunnels in the clay. These tunnels loosen up the soil and let water and air get down into the soil.

4 When digging, a fork should be used instead of a shovel or spade. Why? If you do happen to cut an earthworm in two, it usually will not die but will grow a new head or tail and continue to work in the clay.

5 The earthworm does not like much light or heat.
(a) If an earthworm stays too long in sunlight, it will dry up and die.
(b) During the day it usually stays under the ground, where it will be moist and cool.

6 The earthworm is a great source of food for many creatures. Can you name some creatures that feed on worms?

7 In winter time, the earthworm travels deep down into the soil. Why does it do this?
When springtime comes and the soil begins to warm up, the earthworm starts to burrow through the soil again. It will have a busy time until the winter returns!

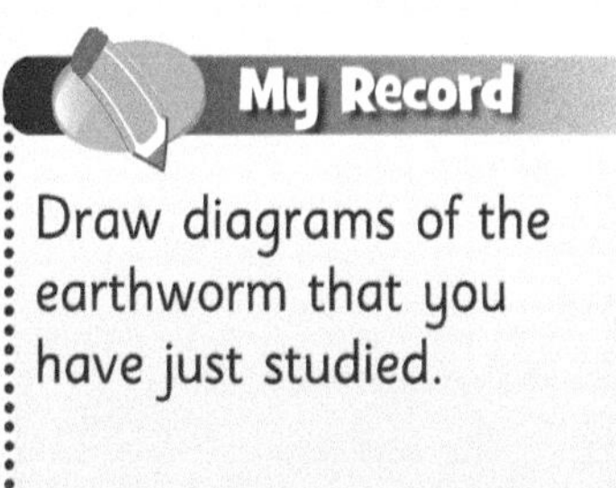

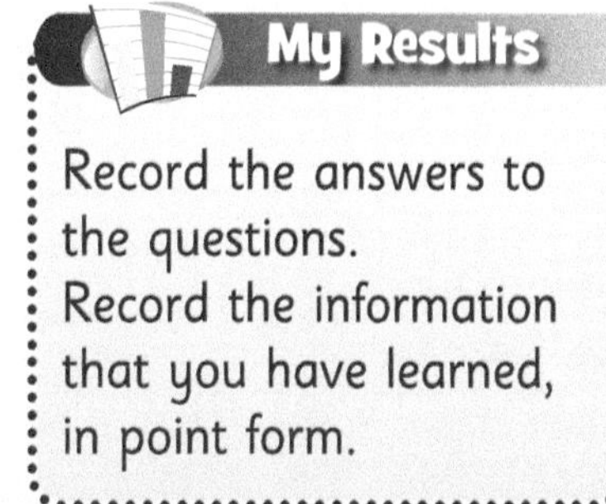

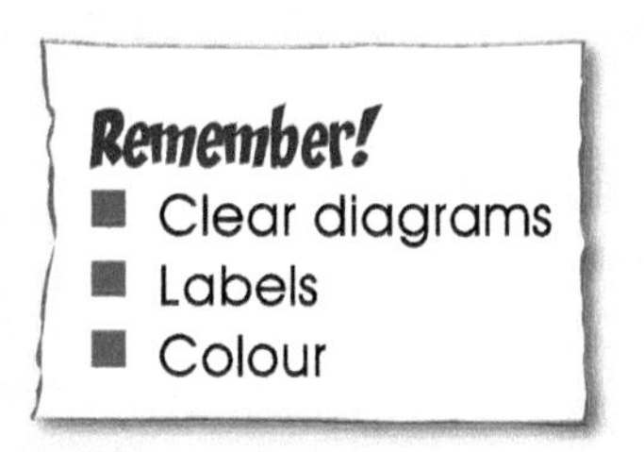

Finding Out More

You will need:
- A container with lid
- Clay
- Stones
- Sand
- A magnifying glass
- Green leaves
- Black sugar paper or cloth

In order to see the earthworm and the work that it does, you can make a wormery.

1. Get a container like the one in the diagram. Line the bottom with a few stones.
2. Put in a layer of damp clay, then a layer of sand.
   Put in another layer of damp clay and another layer of sand.
   Keep repeating this until the container is almost full.
   Put a few green leaves on top.

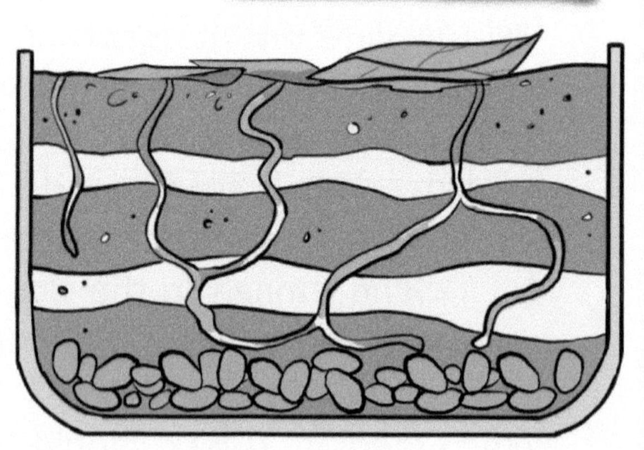

A wormery

3. Go to the garden and find a few worms. Four or five will be plenty.
   Put the worms into the container.
4. Place a lid on the container, and make sure that you put a few holes in the lid.
   Why do you need to put a few holes in the lid?
   Why does the clay need to be moist?
5. Cover the container with black sugar paper or with a cloth. Store it in a cool place.
   Leave it for a week. Then take it back to the classroom.
   Remove the cover. What has happened?
6. Why is the earthworm often called 'nature's ploughman'?

My Record

Draw diagrams of the wormery.

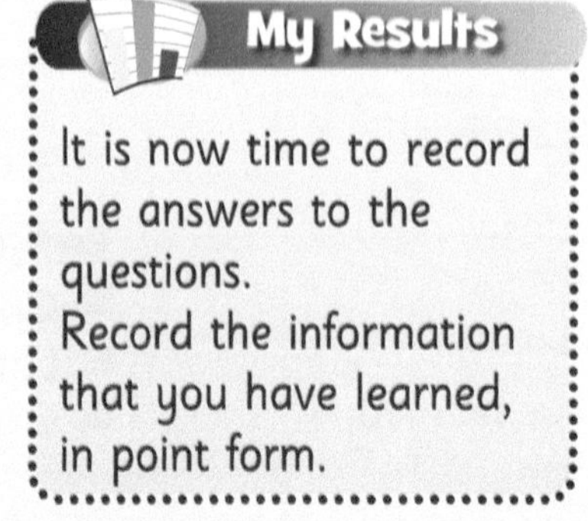

My Results

It is now time to record the answers to the questions.
Record the information that you have learned, in point form.

Remember!
- Clear diagrams
- Labels
- Colour

Detective Time

You will need:
- A sheet of paper
- A magnifying glass

1 Before you return the earthworms to where you found them, you can have a closer look.

2 Carefully take an earthworm and place it on a sheet of paper.
- (a) Watch it move. You will see the muscular action as it moves forward.
- (b) Listen carefully. What do you hear?
- (c) Can you see the bristles underneath the earthworm?
- (d) Get a magnifying glass. Look at the earthworm. What do you see?
- (e) Find the saddle. What does it look like?

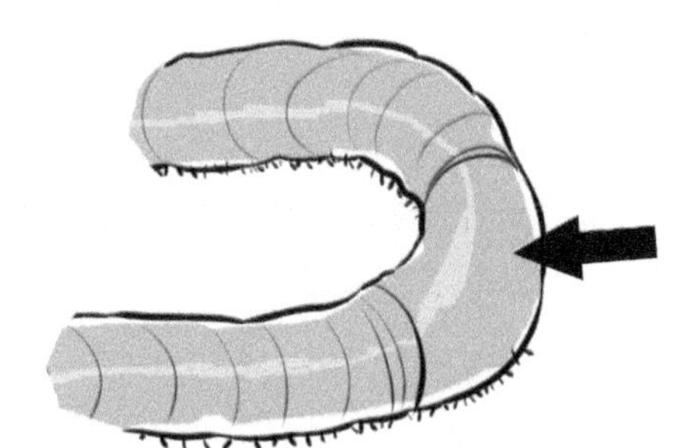

3 Do not keep the earthworm out of the clay for too long, as it will dry up and die!
Now carefully return the earthworm to where you found it.

My Record

Draw diagrams of the earthworm. Show the saddle.
Show the earthworm moving.

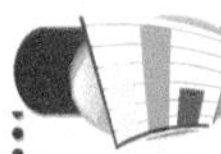

My Results

It is now time to record your results.

Science and the environment

- Composting is a very good way to recycle much of our household waste. Vegetable peels, grass and cardboard are some of the items that can be made into **compost** in a composter. Earthworms help to mix up the different items in the composter. After a while, the waste becomes good compost. This is a natural way of dealing with waste.

Web Watch!
If you would like to find out more about the earthworm, go to this website:
http://www.backyardnature.net/earthwrm.htm

Integration Project

Nature's Ploughman

English

1 What does it mean to be a bookworm? Are you one?
2 Design a book survey sheet to find out what kinds of books your classmates read.
3 Design a book review sheet for your class and give it to your partner.

Gaeilge

Déan péist na bhfocal le rudaí atá le feiceáil sa ghairdín.

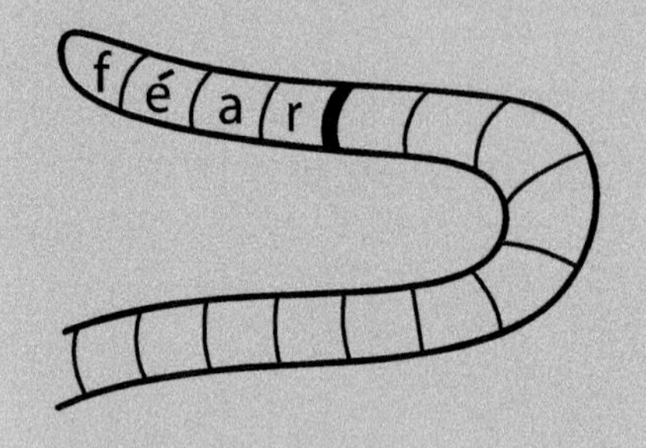

Mathematics

If a worm turned over soil at a rate of 10 grammes per minute,
(a) How much soil would it turn over in an hour?
(b) How much soil would it turn over in 24 hours?
(c) How much soil would four worms turn over in 5 hours?

Music

Did a song ever get stuck in your head all day and it just wouldn't go away? This is called a music worm!! Find out if there was any music worm in your class today.

History

1 Earthworms have survived for millions of years because of their ability to adapt so well to their environment. How do we, human beings, adapt to the environment so that we can survive?
2 Compare farming methods from long ago with modern farming methods.

SPHE

1 How do earthworms benefit (a) farmers, (b) gardeners and (c) fishermen?
2 It is said that earthworms eat their own body weight of food per day. Make up your diet for the day, with the aim of eating your own body weight!

Geography

1 How can we protect insects, while we study them?
2 How do earthworms help us?
3 List the things that you can put into a food composter.

Science

Earthworms and minibeasts in general have an important role to play as part of food chains.
Make up three different food chains.

Does he have teeth?

Yes, he has because he bites when he is in a fight.

Of course he has. He uses them to chew his food.

I think that all creatures that eat food must have teeth.

No, he hasn't because he is a herbivore and doesn't need them.

Identify the main question you need to answer.
Record it now on your Investigation Record Sheet.

Use your Record Sheet to investigate!
My Question
My Prediction
Let's Investigate!
My Record
My Results
My Conclusion

My Prediction

Complete this sentence on your Investigation Record Sheet:

I think that we have ______________________________

____

In Chapter 5 we learned about the different types of food we need to eat. The first part of our body that we use to help us eat food is our teeth.

1. When babies are born, do they have teeth? Explain your answer.
2. At what age do babies begin to grow teeth?
   By the age of three, babies usually have all of their teeth.
   What are these teeth called?
   How many teeth does a young child have?
3. At what age do these teeth begin to fall out?
4. The second set of teeth that we grow should last us for the rest of our lives if we take good care of them. An adult usually has 32 teeth.

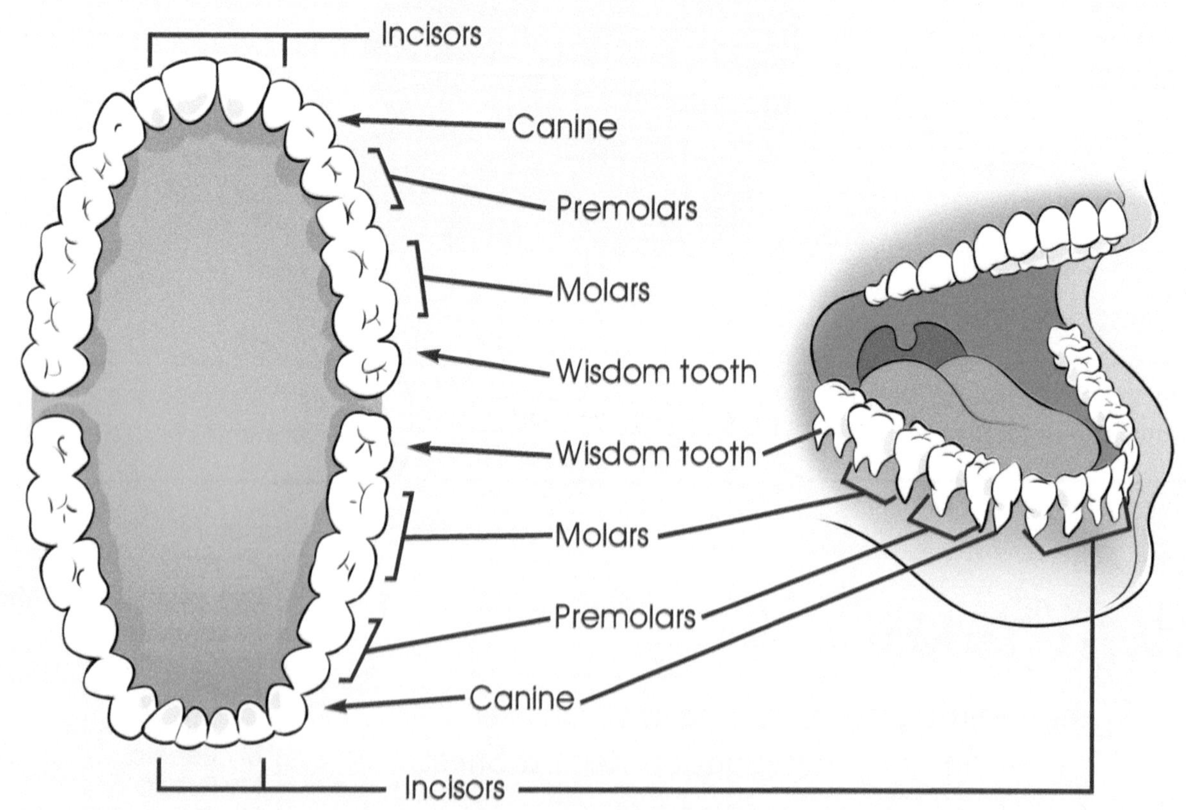

5. Look carefully at the diagrams of our teeth.

   (a) There are eight sharp teeth at the front of your mouth (four on the top and four on the bottom). These are called **incisors**. What job do they do for us?

(b) Beside the incisors are four **canines** (two on the top and two on the bottom).
What job do they do for us? In some creatures the canines are very large. Why is this?
Name a few creatures that need large canines.

(c) Behind the canines you will find eight **premolars** (four on the bottom and four on the top).

(d) Behind them are eight **molars** (four on the bottom and four on the top).
What job do they do for us? Why do we need to crush, grind and chew our food?

(e) When adults think that they are finished growing teeth, four extra teeth grow behind the molars! These are called **wisdom** teeth. People don't grow their wisdom teeth until they are teenagers or older.

My Record

Draw diagrams of your bottom row of teeth.

My Results

Record the answers to the questions. Record the information that you have learned, in point form.

Finding Out More

Run your tongue along your teeth.

1. Can you feel the incisors, both on the top and on the bottom?
2. Can you find the canine teeth? What do you notice about them?
3. Next, see if you can find the premolars and molars. What do you notice about the shape and size of them?
4. Can you find your wisdom teeth? Explain your answer.

Remember!

- Clear diagrams
- Labels
- Colour

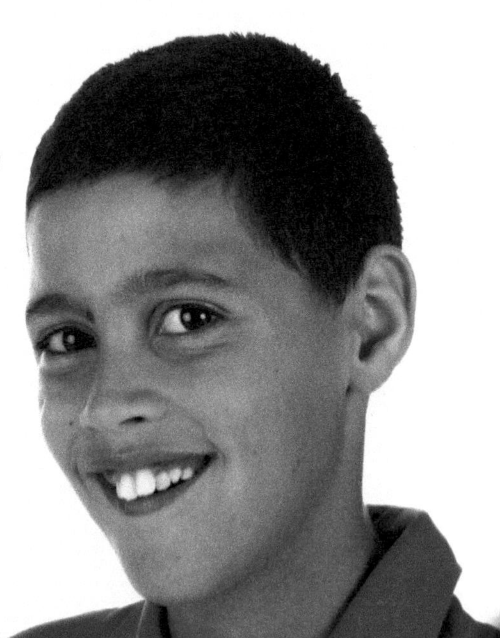

My Record

Draw diagrams of the top row of teeth.

My Results

Record the answers to the questions.

Detective Time

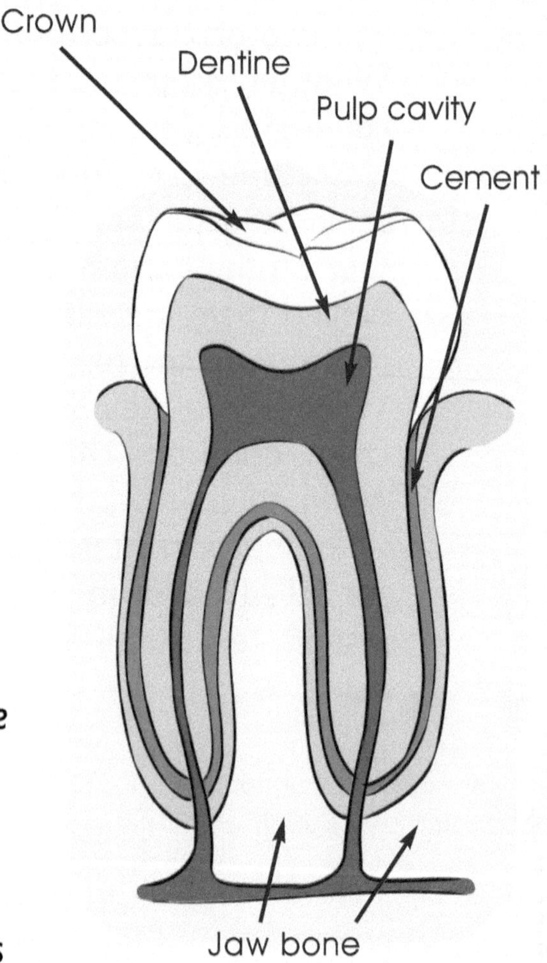

When you look into your mouth, you can see your teeth, but what you see is only a small part of your tooth.

Look carefully at the diagram and learn a lot more about your tooth.

1. The white part of the tooth that you can see is called the **crown**.
   It is made of enamel, which is a hard substance. Enamel protects the tooth.

2. Underneath the crown, there is a substance which is like bone. It is called **dentine**. Dentine is softer than enamel.

3. The **pulp cavity** contains blood vessels and nerves. When we get a toothache, it is the nerve that causes the pain. The blood vessels bring food and oxygen to the tooth.

4. The tooth fits over the jaw bone and is held in place and protected by **cement**. The cement is a bone-like substance and is very strong.

5. It is very important to take good care of our teeth.
   Remember they have to survive for the rest of our lives.
   How often should you brush your teeth?

6 We all have bacteria in our mouths.

(a) Bacteria gather around our teeth and form a coating called plaque.

(b) When we eat foods with sugar in them, the sugar reacts with the plaque and forms an acid.

(c) Acid is strong and begins to wear away the enamel on the crowns of our teeth.

(d) This leads to decay.

(e) We need to brush our teeth to get rid of as much plaque as possible, in order to prevent the build-up of acid.

7 Name some of the sugary foods that you eat.

8 What should you do after eating sugary foods? Why?

Draw diagrams of a tooth.

Record the answers to the questions.
Record the information that you have learned, in point form.

- Clear diagrams
- Labels
- Colour

Science and the environment

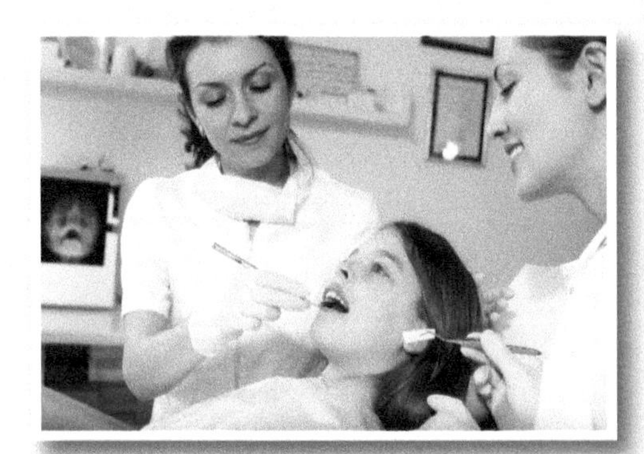

- Dentists help us with the care of our teeth. They do fillings, teeth extractions, crowns and even clean our teeth.
- Toothpaste and other products have been invented to help us look after our teeth.
- Surgeons can do amazing work to help people who have damaged their teeth or their jaw bones in an accident. They can reconstruct jaw bones, teeth and even the skull, if a patient needs it.

Web Watch!
If you would like to find out more about teeth, go to this website:
http://kidshealth.org/kid/htbw/teeth.html

Integration Project

English

1 Read the funny poem 'Oh, I wish I looked after me teeth' by Pam Ayres:
http://www.poetryarchive.org/poetryarchive/singlepoem.do?peomId=11736

2 Rewrite the poem using the correct grammar.

3 Write about whether you would/would not like to be a dentist when you are older.

Gaeilge

1 Tomhais:
Ollphéist dearg
Istigh sa phluais
Is clocha bána
Thíos is thuas.

2 Scríobh scéal 'Ag dul go dtí an bhfiaclóir'.

Mathematics

1 How many teeth do adults have?

2 How many teeth do children have?

3 How many teeth are there altogether in your family?

4 If a filling costs €60 and an extraction costs €80, how much would you have to pay the dentist for 2 fillings and an extraction?

Say Cheese!

Music

Explore the sounds that teeth make.

History

The first bristle toothbrushes, similar to what we use today, were used in China around 1498.
Find out more about the history of looking after our teeth on:
http://www.thehistoryof.net/the-history-of-teeth-whitening.html

SPHE

Learn more about keeping your teeth healthy by visiting this website:
http://www.healthyteeth.org/

Geography

Write out the directions or draw a map of how you would get to the nearest dentist in your area.

Science

1 Make lists of foods that we (a) chew, (b) bite and (c) crunch.

2 Name the type of teeth involved in breaking down each of these foods.

Will all substances dissolve in water?

Identify the main question you need to answer.
Record it now on your Investigation Record Sheet.

My Prediction

Complete this sentence on your Investigation Record Sheet:

I think that the substances that we have chosen

____

Let's Investigate!

You will need:
- Sand, rice, salt, pepper, sugar, flour
- Six containers
- A thermometer
- A spoon
- Cold water

1. Get six containers and put some cold water into each.
   Get a thermometer and measure the temperature of the water.
2. Put sand in the first container, rice in the second container, salt in the third one, pepper in the fourth one, sugar in the fifth one and flour in the sixth one.

3. Leave for a little while. What do you notice?
   You could use a table like the one below to record your findings.
4. Get a spoon and gently stir each container for 30 seconds.
   Leave for a little while. What do you notice?
5. Does stirring improve dissolving?

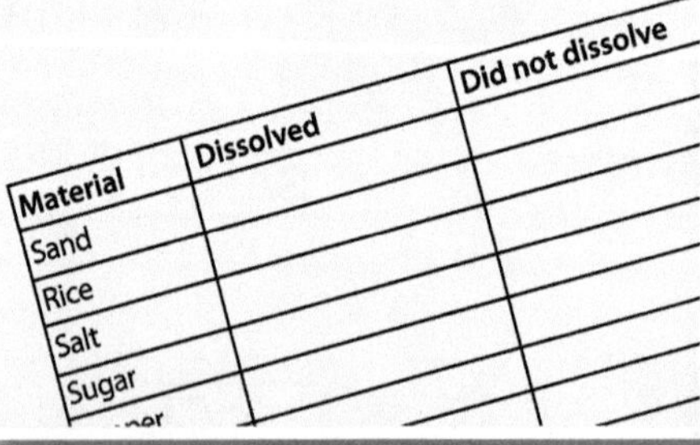

Material	Dissolved	Did not dissolve
Sand		
Rice		
Salt		
Sugar		

My Record

On your Investigation Record Sheet, draw diagrams of your investigation.

My Results

It is now time to record your results. Use a table like the one above.

My Conclusion

Now that your class has discussed the investigation, you can record the conclusion.

Finding Out More

We are now going to do the same experiment, but this time we will use warm water.
Be very careful when using warm water.
Your teacher will give you the warm water and will make sure that it is not too hot.

You will need:
- Sand, rice, salt, pepper, sugar, flour
- Six containers
- Warm water
- A thermometer
- A spoon

1. Once again, get six clean containers. Get a thermometer and record the temperature of the water.
2. Put some warm water in each jar and add a little of the different substances as before. Leave for a little while. What do you notice? You could use the same table as before to record your results.
3. Get a spoon and gently stir each container for 30 seconds. Leave for a little while. What do you notice?
4. Does stirring improve dissolving?
5. Are the results any different when you use warm water?
6. You could now choose different substances and experiment with them to see if they will dissolve in warm or cold water.

Record your diagrams, results and conclusion on your Investigation Record Sheet, just as you did for the Let's Investigate! exercises.

Detective Time

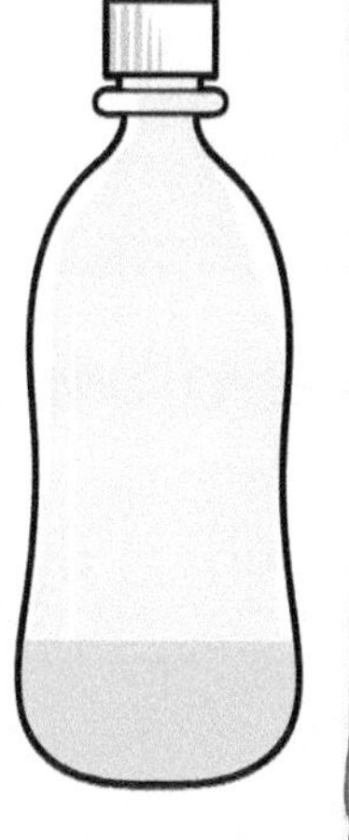

You will need:

- Cooking oil
- Syrup
- Food colouring
- A plastic bottle
- A jam jar
- A piece of cork
- A piece of a candle
- A small grape
- Warm water
- Cold water

So far, we have experimented with different substances but all of them were solids of one type or another.
Now let's experiment with different liquids.

1. Get a plastic bottle.
   (a) Pour some cooking oil into the bottle until it is about 3cm high within the bottle.
   (b) Get some cold water and add a little food colouring to it.
   (c) Pour the water into the bottle and see what happens.
   (d) Repeat the experiment, using warm water this time.
   (e) What happened?
   (f) Take one of the bottles containing the water and the oil.
   (g) Put the lid on it and give it a good shake. What happened?
   (h) Leave it to one side for a while and see what happens.

2 Get a jam jar.

(a) Pour some syrup into it until it is about 2cm high within the jar.
(b) Get some cooking oil and some coloured water.
(c) Pour about 2cm of cooking oil on top of the syrup. What happens?
(d) Now pour in about 2cm of the coloured water.
(e) Observe carefully. What happens?
(f) Get a small piece of cork, a little wax from a candle and a small grape.
(g) Put each item into the jam jar.
(h) Observe carefully. What do you notice?

3 What happened to the bottle with the oil and the water that you shook earlier?

My Record

On your Investigation Record Sheet, draw diagrams of your investigations.

My Results

It is now time to record your results.

My Conclusion

Now that your class has discussed the investigation, you can record the conclusion.

Science and the environment

Scientists have shown us that we get new products by mixing substances:

- Concrete is made by mixing water, sand and cement.
- Many foods are dried but need to be mixed with water to get them ready for eating. They are dried to help preserve them. Gravy, soup, porridge and pancake mix are a few examples. Can you think of more?

Web Watch!
If you would like to find out more about solutions in water, go to this website:
http://www.chem4kids.com/files/matter_solution.html

Integration Project

English

1. When cooking, we often have to dissolve something in water. Can you think of examples?
2. Write out the instructions for making jelly.

Gaeilge

1. Scríobh scéal beag le teidil 'Ag Snámh' nó 'Lá ar an Trá'.
2. Cén caitheamh aimsire is fearr leat?

Mathematics

1. If I had a 1-litre container of water to fill smaller containers with a capacity of 200ml each, how many of the smaller containers would I fill?
2. What fraction of 2 litres is 500ml?

Water Wizardry

Science

Hot water dissolves better than cold water because it has faster-moving molecules, which are spread further apart than the molecules in the cold water. With bigger gaps between the molecules in the hot water, more sugar molecules can fit in between them.

Check out how many sugar cubes will dissolve in a glass of cold water and in the same amount of hot water.

History

Drying is the oldest method of preserving food. Find out more on:
http://www.ssrsi.org/sr1/Cook/drying.htm

Geography

Do a project on the water cycle. Check out this website:
http://www.kidzone.ws/water/

SPHE

1. Check out food packaging which says 'Just add water'.
2. What advantages do products like these have over other products?

16 Flower Power

What is a weed?

Identify the main question you need to answer.
Record it now on your Investigation Record Sheet.

My Prediction

Complete this sentence on your Investigation Record Sheet:

I think that the flower in the picture on page 83 is

____

Let's Investigate!

1 Take a good look at the pictures of this common wild flower. Where have you seen it before?

2 This plant is a perennial. What does that mean?

3 It flowers from March to October. It is a native Irish plant. It has a bright yellow flower and this turns into an unusual type of seed head. The seed head is often called a clock. How does this plant disperse its seeds?

to disperse: to scatter

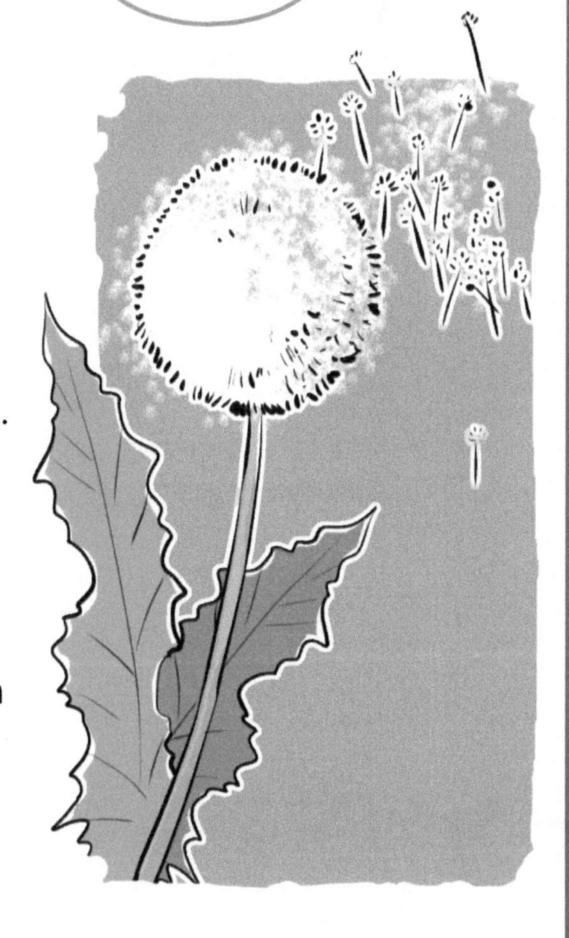

4 The leaves remain low down at the base of the plant. Look carefully at the diagram. Describe the leaves.

5 A long stem grows up from the leaves and the flower appears on the top. The plant grows to between 10cm and 50cm in height. The stem contains a milky substance called latex. The plant has a long root that grows deep down into the ground. Why is this, do you think?

6 Some people gather the leaves and use them in salads. The flowers are also used to make homemade wine.

7 Can you name this plant?

My Record

On your Investigation Record Sheet, draw diagrams of the flower that you have just studied.

My Results

It is now time to record the answers to the questions. Record the information that you have learned, in point form.

Finding Out More

1 This is another common wild flower.
Look carefully at the pictures.
Have you ever seen a flower like it before?
Where?

2 This plant is a perennial. What does that mean?

3 It flowers from May to August. It is a native Irish plant.
It has a golden yellow flower.
It has five petals with the flower head in the centre.

4 The leaves often remain close to the ground.
How many segments are in each leaf?

5 The plant sends out runners, which take root in the clay and so a new plant is formed.
How is this different from how most other plants reproduce?

6 A long stem grows up from the leaves and the flower head is formed on top of the stem.
It can grow to 60cm in height.

7 Can you name this plant?

My Record

Draw diagrams of the flower that you have just studied.

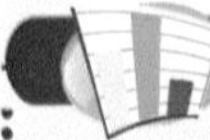

My Results

It is now time to record the answers to the questions.
Record the information that you have learned, in point form.

Remember!
- Clear diagrams
- Labels
- Colour

Detective Time

1. This is another common wild flower. Look carefully at the pictures. Have you ever seen this flower? Where?

2. This plant is a perennial.

3. It flowers from December to May. It is unusual to have a flower that comes into flower in the winter period. This feature makes this wild flower all the more special. It is a native Irish plant.

4. It has a pale yellow flower. How many petals are in each flower? What is unusual about each petal?

5. The leaves grow low down but are often more raised than those of either the dandelion or the buttercup. Describe the leaves.

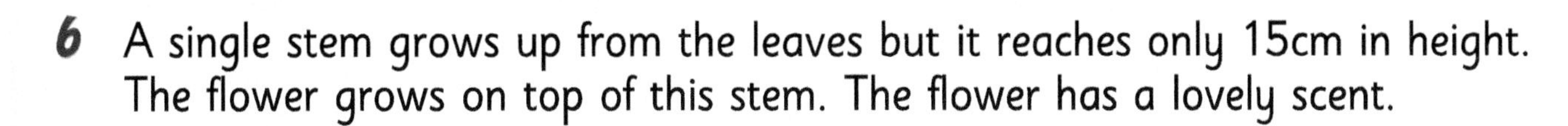

6. A single stem grows up from the leaves but it reaches only 15cm in height. The flower grows on top of this stem. The flower has a lovely scent.

7. You will often see this plant growing in woods, under hedgerows, in meadows and along the roadside. Some people like to put the flowers into salads.

8. Can you name this plant?

My Record

Draw diagrams of the flower that you have just studied.

My Results

It is now time to record the answers to the questions.
Record what you have learned, in point form.

Remember!

- Clear diagrams
- Labels
- Colour

Make and Do

A Flower Press

In the month of May it may be possible to go on a short walk from your school in order to see these flowers. All three of the flowers should be still available in late April or early May. Your teacher will have chosen the site for the walk. *Be very careful when walking on the road.* Bring some paper and a pencil, so that you can sketch the flowers in the wild.

If you wish, you can bring back samples of all three flowers to the classroom, to examine them more closely.

When you are finished examining them, you could dry and press the plants.

How to make a flower press

1. Get two pieces of stiff card. They would need to be about 30cm by 30cm.
2. Cut approximately 8 pieces of newspaper of the same size.
3. Place one piece of card on your table.
   - Put 3 or 4 sheets of newspaper on the card.
   - Place the flowers you want to press on top of this paper.
   - Put 3 or 4 sheets of paper on top of the flowers.
   - Place the second sheet of card on the top.
   - Put some weight on the top sheet of card.
4. Leave the flower press for a week or so and by then the flowers should be dried and pressed.
5. When you take the press apart, you can use the flowers to form displays on sheets of card or to make greeting cards.
6. Don't forget to label each flower if you use them in a display.

Very Interesting

Science and the environment

- Scientists have discovered and named more than 350,000 different species (types) of plants. They have divided all plants into two groups:
  - those that produce a flower
  - those that do not produce a flower.

  More than two-thirds of all plants produce flowers.

- Grass is a flowering plant. It is a great source of food for animals but is also a great source of food for humans. Wheat, corn and rice are all types of grass and they produce most of the world's food.

- Plants are used in the manufacture of many medicines.
  - The common foxglove (digitalis) is used in the manufacture of medicines to treat heart disease.
  - The bark of the willow tree was once used to make aspirin.
  - Herbalists use several herbs and plants to treat patients. Herbal teas are very popular.
  - Herbal medicine is very commonly used in China, India and Russia.

- *Never eat leaves, berries or herbs without checking with an adult first.*

Web Watch!
If you would like to find out more about wild flowers and their Irish names, go to this website:
http://www.irishwildflowers.ie/

The foxglove

Rice harvest

Herbal tea

Integration Project

English

1 When exploring nature 'take nothing but photographs; leave nothing but footprints'. Discuss.
2 Compose a poem about wild flowers.

Gaeilge

1 Cén Gaeilge atá ar 'dandelion', 'primrose', 'buttercup'?
2 An bhfuil ainmneacha bláthanna eile ar eolas agat? Céard iad?
3 Scríobh scéal gearr faoi lá a cheannaigh tú bláthanna do do Mhamaí, nuair a bhí sí tinn.

Mathematics

Carry out a survey in your group to find out everyone's favourite wild flower.
Put the results on a block chart and make up your own questions.

Music

Compose a rap about wild flowers.

Flower Power

History

Find out more about the usefulness of dandelions throughout history.
Go to:
http://www.innvista.com/health/herbs/dandelio.htm

Visual Arts

Make cards and stationery using pressed wild flowers. Add a greeting such as 'Thank you' or 'Happy Birthday'.

Geography

1 Are all weeds welcome in gardens or are some more welcome than others? Discuss.
2 Do you think that primroses, buttercups and dandelions are weeds and should we use weed killer to get rid of them? Why/why not?

Science

List ways in which wild flower seeds are dispersed.

What will happen to these two pages when I blow between them?

Identify the main question you need to answer.
Record it now on your Investigation Record Sheet.

My Prediction

Complete this sentence on your Investigation Record Sheet:

I think that the pages will ______________________________

____

You will need:
- 2 sheets of A4 paper
- 2 sheets of A4 card

Let's Investigate!

1. Begin by getting two A4 sheets of paper. Hold the pages as in the diagram. Blow between the pages. What happened?
2. Try blowing a little harder. What was the result this time?
3. Talk to your friends. Can you do anything to make those pages go apart?
4. Try it out. Did it work?
5. Get two sheets of card. Will the result be the same this time?
6. Blow between the two sheets of card. What happened?

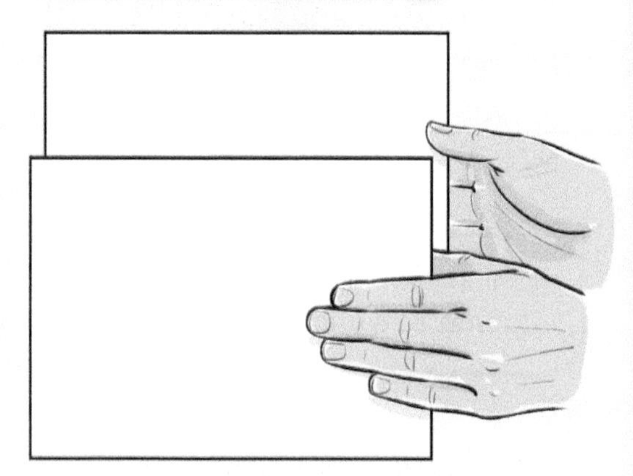

My Record

On your Investigation Record Sheet, draw diagrams of your investigation.

My Results

It is now time to record your results.

My Conclusion

Now that your class has discussed the investigation, you can record the conclusion.

Finding Out More

1. Get an A4 sheet of paper and fold it as in the diagram. Make sure that you hold on to the page.
2. Blow *under* the shape. What do you notice?
3. Where would you need to blow so that the shape would rise up slightly? Try it out. Did it work?

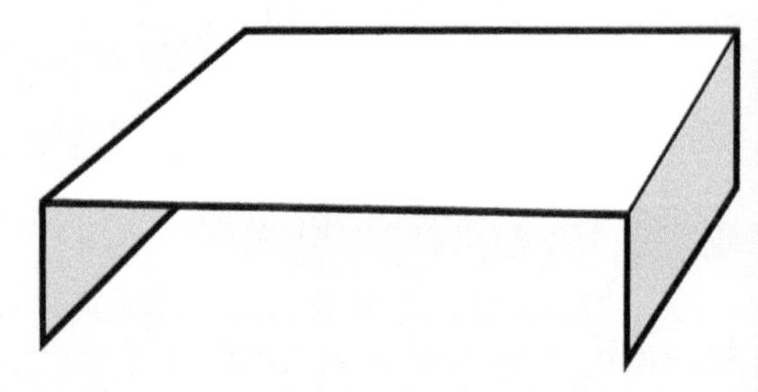

Record your diagrams, results and conclusion on your Investigation Record Sheet, just as you did for the Let's Investigate! exercises.

Detective Time

You will need:
- A funnel
- A table tennis ball

1. Get a funnel and a table tennis ball. Hold the funnel with the wide side upwards and put the ball into it, as in the diagram.
   What do you think will happen to the ball if you blow hard through the spout of the funnel while keeping it in an upright position?
2. Try it out. What happened?
3. If you were to repeat the experiment and, while still blowing hard, turned the funnel towards the floor, what do you think would happen to the ball?
4. Try it out a few times. What happened?

Record your diagrams, results and conclusion on your Investigation Record Sheet, just as you did for the Let's Investigate! exercises.

Remember!
- Clear diagrams
- Labels
- Colour

Science and the environment

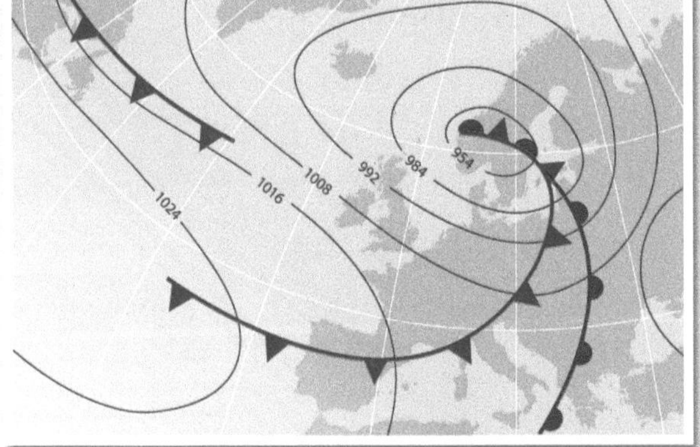

- Scientists who study weather and weather forecasting are called meteorologists. They are able to predict what the weather will be like in the immediate future.
- Within our atmosphere there are areas of high pressure and areas of low pressure. Wind is moving air. Wind moves from areas of high pressure to areas of low pressure. The greater the difference in pressure, the faster the air moves. Storms are created by air moving from areas of very high pressure to areas of very low pressure.

Web Watch!
Learn more about wind energy on:
http://tonto.eia.doe.gov/kids/energy.cfm?page=wind

Integration Project

Moving Air

English

List the advantages and disadvantages of travelling by air. Compare and discuss your results with a partner.

Visual Arts

Design your own flying machine. Draw it and label any special features.

Physical Education

Play a game of air football, while sitting at your table opposite your partner.
Roll up a small piece of paper to make a ball and mark off goal areas. Each of you attempts to blow the 'ball' into the opposite goal.

SPHE

1. Check out your lung capacity: Hold your breath and, using a stopwatch, time how long you can keep it in.
2. See who in the class can hold his or her breath the longest.

Gaeilge

Déan liosta de na rudaí gur féidir leo eitilt sa spéir.

Geography

1. List the ways in which human beings pollute the air.
2. How do we use air positively?

Mathematics

Drop a sheet of paper from different heights measured in cm and measure in seconds the time it takes to reach the floor. Work out the speed of the paper falling in cm per second. Compare your findings with those of your friend.

History

Describe the ways in which air travel has changed our lives, under the following headings: Economy, Culture, Food, Education, Social.

Science

Make a list of the ways in which we travel by air.

Why do plants grow?

Identify the main question you need to answer.
Record it now on your Investigation Record Sheet.

My Prediction

Complete this sentence on your Investigation Record Sheet:

I think that we can eat ______________________________
____

Let's Investigate!

Begin by discussing with your group the different parts of common plants that we eat.

1. What are the parts of the common plants?
2. Do we eat roots? If so, which plants?
3. Do we eat stems? If so, which plants?
4. Do we eat leaves? If so, which plants?
5. Do we eat flowers? If so, which plants?
6. Do we eat seeds? If so, which plants?
7. Do we eat fruits? If so, which plants?

My Record

On your Investigation Record Sheet, draw diagrams of the different parts of plants that we eat.

My Results

It is now time to record your results on the table provided.

Finding Out More

We certainly eat lots of plants, but now it is time to talk with your group about plants that we drink!

1. Name one drink that is made from a plant.
2. With your group, investigate where tea comes from.
   - From what is tea made? How is it made?
   - Examine the countries that produce it. Does it need a special climate in order to grow?
3. Investigate where coffee comes from.
   - From what is it made? How is it made?
   - Where is this found? Examine the countries that produce it.
   - Does it need a special climate in order to grow?
4. Look at the ingredients of a cola drink.
   - What gives it its distinctive flavour?
   - Find out all you can about this ingredient.
   - Does it need a special climate in order to grow?

5 Next, investigate where chocolate and cocoa come from.
- From what are they made?
- How are they made? Where is this found?
- Examine the countries that produce it. Does it need a special climate in order to grow?

My Record

Choose one plant that we use to make a drink. Draw diagrams of it on your Investigation Record Sheet.

My Results

It is now time to record your results. How will you record the new information that you have discovered?

Detective Time

So far in this lesson we have found that we eat and drink plants. Now let's think about plants that we wear.

1 Can you think of any plants that we wear?

2 Investigate where cotton comes from.
- From what is it made? How is it made?
- Where is this found? Examine the countries that produce it.
- What products are made from cotton?
- Does it need a special climate in order to grow?

3 Natural silk comes from a creature that lives on a plant. Investigate how silk is made.
- Where is this found? Examine the countries that produce it.
- Does it need a special climate in order to grow?

4 Linen is made from flax.
- Flax was grown in Northern Ireland during the twentieth century.
- It was used to make ropes and large jute bags as well as linen cloth.

My Record

Choose one plant that we wear. Draw diagrams of it on your Investigation Record Sheet.

My Results

It is now time to record your results. How will you record the new information that you have discovered?

Very Interesting

Science and the environment

- Flax seeds have oil in them called flaxseed oil. This oil is made into linseed oil. It is used in paints and varnish to help preserve wood and leather.
- Linseed oil is high in omega oils. People take it as a food supplement. It is thought to help people with arthritis and to protect us against cholesterol and heart disease.

English

1 Make up your own alphabet of plants that we eat:
http://www.kidcyber.com.au/topics/plantsalpha.HTM
2 Write out instructions on how to make a mug of hot chocolate.

Gaeilge

1 Tomhais: Chomh bán le sneachta; chomh glas le féar; chomh dearg le fuil agus chomh dubh le simléar.
2 Ainmnigh na glasraí agus na torthaí is fearr leat.

Mathematics

Restaurant price list: coffee €2.50, tea €1.75, hot chocolate €2.00, salad sandwiches €2.75, chocolate chip muffins 80c. If your family ordered the following: 2 coffees, 1 tea, 5 salad sandwiches and 6 muffins, what would be the total cost?

Music

Compose and perform a rap about plants that we eat, drink and wear and how we need them to survive.

Geography

Investigate how chocolate is made from the cacao bean.

History

1 Fine linen was used for burial shrouds for the ancient Egyptian Pharaohs.
Find out more about how fabric was used long ago:
http://www.fabriclink.com/university/history.cfm
2 Find out more about the famous Shroud of Turin on:
http://encyclopedia.kids.net.au/page/sh/Shroud_of_Turin

SPHE

Write out a catchy advertisement for your favourite juice cocktail.

Science

We eat many different parts of plants – bulbs, flower buds, fruit, leaves, roots, seeds, stems, stalks and tubers. Name plants that supply us with each of these.
http://www.naturegrid.org.uk/plant/foodparts.html#top

IRELAND
GRAND SLAM CHAMPIONS

RBS 6 NATIONS 2009

GILBERT

IRELAND GRAND SLAM CHAMPIONS

RBS 6 NATIONS 2009

JIM LYONS

First published 2009

Nonsuch Publishing
119 Lower Baggot Street
Dublin 2
Ireland
www.nonsuchireland.com

Frontispiece: Ronan O'Gara (David Maher/Sportsfile)

British Library Cataloguing in Publication Data.
A catalogue record for this book is available from the British Library.

ISBN 978 1 84588 965 4

Cover and internal design by Emma Jackson
Printed by GPS Ltd in Northern Ireland on Forestry Stewardship Council (FSC) certified paper

Contents

Introduction 7

Ireland v France 8

Ireland v Italy 24

Ireland v England 36

Ireland v Scotland 56

Ireland v Wales 68

The Homecoming 88

canterbury
O2
O2
BRAINS
SA
RBS

Introduction

Every fan of rugby in Ireland has heard it said that the team we have today is the Golden Generation of Irish Rugby. In Brian O'Driscoll, we have the best centre in the world. In Paul O'Connell, we have a mountain of a man, twice a Heineken European Cup Winner as Captain of Munster. In Ronan O'Gara we have an out-half who is now the highest points scorer in the history of the 6 Nations, and so much more than that. And in Europe's greatest competition, we have been waiting for sixty-one years for a Grand Slam; sixty-one years since Karl Mullen led Ireland to that kind of victory. Playing in that team was Jack Kyle, eighty-three now, but then an out-half, regarded by many as Ireland's greatest ever player. A modest man, Jack would no doubt be happy about the number of players in 2009 who might now challenge him for that honour. O'Driscoll, O'Connell, O'Gara. Or Rory Best, another leader on the pitch. Or Stringer, the people's favourite. The list is as long as the squad sheet, and longer. It is a Golden Generation, and on Saturday 21 March 2009, in the Millennium Stadium in Cardiff, they showed us why.

No words can adequately describe the story of the last few weeks, so we'll try and show you, with some of the greatest images of the greatest moments of the greatest thing that has happened to Ireland in a long, long time.

The 6 Nations, 2009.

Opposite: Ireland captain Brian O'Driscoll, with captains Sergio Parisse (Italy), Steve Borthwick, (England), Lionel Nattet (France), Mike Blair (Scotland), and Ryan Jones (Wales), at the RBS 6 Nations launch at the Hurlingham Club, London. (David Maher/Sportsfile)

Ireland v France

CROKE PARK, SATURDAY 7 FEBRUARY 2009

So, here it was. All potential, just like the last time, and the time before. Under Eddie O'Sullivan we had come close to greatness, with three Triple Crowns in five years, but this wasn't enough anymore. We might never have a team like this again, and what we wanted was a Grand Slam – the ultimate accolade in Northern Hemisphere rugby. To be truly great, it was not enough to deserve this. Sometimes you need your name carved in silver and remembered forever. We needed this, as people were beginning to say, as a country.

France are a spectacular rugby nation, and we had lost the last six games against them. In Croke Park, in 2007, we thought it was all over until it was snatched away in the dying seconds. But now they were back, on the 100th anniversary of the teams' first meeting, with some familiar faces – and, in Chabal, one of the most recognisable heads in world rugby.

This was Declan Kidney's first 6 Nations as manager of the Irish team, coming fresh from a European Cup win with Munster. Today there was no Peter Stringer, who had been the scrum-half for Ireland since 2000, amassing eighty-five caps along the way. His Munster teammate Tomas O'Leary would partner Ronan O'Gara. Paddy Wallace of Ulster was in the centre, and Stephen Ferris was in the back row.

An incredible, free-flowing game of rugby began, and O'Gara took the first points with a penalty. But Harinordoquy countered with a try in the corner. It took Jamie Heaslip to break the line for Ireland. A fantastic try. At half time, the score was 13-10 to Ireland.

The second half opened, and within minutes … O'Driscoll, flashing through the French defence just like he used to do almost a decade earlier. He was touching the ball down while a lot of the crowd were still making their way back to their seats. Stunning try. At 20-10, it was looking good, but then a timely chip, and France were back in it with a try from Medard. And, minutes later, a drop goal for France. 20-18. Things were getting tense in the stands now. Paul O'Connell put in a huge tackle. 'Come on, you big, beautiful man', roared someone from the crowd. Who was going to settle this? Gordon D'Arcy, it would seem. For a player just back from a lengthy injury, D'Arcy showed incredible strength to get over the line, and put Ireland clear. But a French penalty got them back in range. With two minutes left on the clock, however, O'Gara put the ball, and the French, away with a penalty. Ireland win 30-21.

Ireland supporters, from left, Barry Kearney, Jonathan Callaghan, Dermot McGonagle and James Cooke, all from Carndonagh, Co. Donegal, pictured on O'Connell Street ahead of the game. (Stephen McCarthy/Sportsfile)

France supporters ahead of the game. (Stephen McCarthy/Sportsfile)

GUINNESS
MERCER
GUINNESS
GUINNESS
Ulster B
GUINNESS GUINNESS GUINNESS GUINNESS
Enjoy GUINNESS Sensibly

Above: A French fan sings the national anthem, 'La Marseillaise' before the game. (Brendan Moran/Sportsfile)

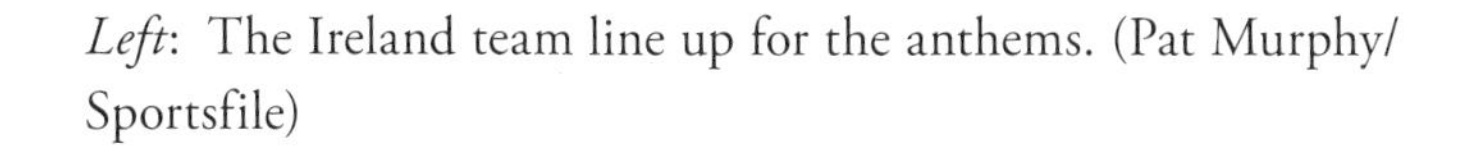

Left: The Ireland team line up for the anthems. (Pat Murphy/ Sportsfile)

Left: Ronan O'Gara. (Brendan Moran/Sportsfile)

Opposite: Sebastien Chabal. (Brendan Moran/Sportsfile)

Opposite: Jamie Heaslip on his way to scoring that try. (Matt Browne/Sportsfile)

Right: Stephen Ferris takes the ball in the lineout against Lionel Nallet. (Matt Browne/Sportsfile)

Above: Paul O'Connell prepares to take a hit. (Pat Murphy/Sportsfile)

Opposite: Paddy Wallace, already showing the signs of just how physical this match was. (Brendan Moran/Sportsfile)

10
GILBERT

The other side of O'Gara. A tackle from Florian Fritz shows it's not all about kicking for the out-half. (Brendan Moran/Sportsfile)

Brian O'Driscoll escapes the attentions of Lionel Beauxis and Yannick Jauzion. (Pat Murphy/Sportsfile)

Above: Gordon D'Arcy celebrates his try against France with Tommy Bowe, Brian O'Driscoll, Tomas O'Leary and Paul O'Connell. (Matt Browne/Sportsfile)

Opposite: Brian O'Driscoll is congratulated by Paddy Wallace, left, and Rob Kearney after his try. (Matt Browne/Sportsfile)

FRANCE

Ireland v Italy

STADIO FLAMINIO, SUNDAY 15 FEBRUARY 2009

A rousing welcome in the Italian sunshine preceded this game, but there was a lot at stake for both teams. While Ireland had been getting back to their best against France, the Italians had been beaten badly by England, and mauled to an even greater degree by the press. The good feeling of the opening ceremony was consequentially short-lived and, within the first minute, Andrei Masi took Rob Kearney out with a high and dangerous tackle. Once the furious teams had been pulled apart, a yellow card followed, and Italy were down to fourteen men. This should have helped, but after two Italian penalties, Ireland were 6-0 down. Plenty of time left, maybe, but people were repeating that a little too often now, almost willing it to be true. But an intercept from Tommy Bowe, and a scintillating run most of the length of the pitch, and we were back in the lead. But not for long … another Italian penalty put them 9-7 ahead. And then Ronan O'Gara was sin-binned for a despairing tackle, which the impartial observer might probably say was off the ball. The referee certainly did, and now we were the ones with fourteen men. But cool heads prevailed, and after a patient build up, going through some nineteen phases, Stephen Ferris handed off to Luke Fitzgerald for his first try for Ireland. The world was getting back on its axis, and Ireland went into the break 14-9 up.

The second half opened, and Ireland began pounding the Italian line; before long David Wallace showed some great feet to weave through for another score. O'Gara, his sentence served, added the points, and followed it with another penalty. Things were getting a bit more comfortable now, at 24-9.

With only a few minutes on the clock, D'Arcy and Fitzgerald combined brilliantly after a quick line-out to add another try. That should have probably been it, but moments later Brian O'Driscoll intercepted another Italian pass, and, from even further out than Bowe had been earlier, blazed home to neatly put a gloss on the result. Ireland win 38-9. Bring on the English.

The Italians put on a colourful performance before the game. (Brendan Moran/Sportsfile)

Irish rugby fans, from left, Karen McArdle, Katie MacCarthy, Julie Tynan, Kate Pierse, Justine Byrne, Annita McGettidan, Isabella O'Keeffe and Catherine Forsyth before the game. (Matt Browne/Sportsfile)

Rob Kearney in full stride. In the opening minutes, opposing full-back Masi sought to break Kearney's rhythm with a high tackle, but to no avail. (Brendan Moran/Sportsfile)

Above: Stephen Ferris breaks through the tackle of Luke McLean, Italy. (Brendan Moran/Sportsfile)

Opposite: Tommy Bowe is grappled by Matteo Pratichetti. (Brendan Moran/Sportsfile)

Tommy Bowe races through the Italian defence on his way to scoring Ireland's first try. (Brendan Moran/Sportsfile)

Denis Leamy is sandwiched by Italian defenders. A panini?
(Brendan Moran/Sportsfile)

Luke Fitzgerald goes over for the second try of the game. (Matt Browne/Sportsfile)

David Wallace touches down for Ireland's third try at the opening of the second half. (Brendan Moran/Sportsfile)

Paddy Wallace leaves the pitch with a blood injury in the company of team doctor Dr Gary O'Driscoll.
(Brendan Moran/Sportsfile)

Brian O'Driscoll on his way to score his side's fifth try of the game. Paul Griffen looks about ready to go home. (Matt Browne/Sportsfile)

Ireland v England

CROKE PARK, 28 FEBRUARY 2009

If ever there was a game that doesn't need an introduction, that doesn't need to be hyped up to get the crowd going, it is Ireland against England. In Croke Park. If you could add some poignancy to this fixture, it might be that this was the last time Ireland is likely to play England in this ground, as the revamped Lansdowne Road is due to host the next clash in 2011. The last time we played here Ireland ran out exultant 43-13 winners, but no one could take anything for granted in this game. Anyone lucky enough to be in the ground for the anthems knows how electric an atmosphere can get.

Ireland fielded an unchanged team, for the third successive time. Paddy Wallace had taken some savage hits in the last games, but was still keeping Gordon D'Arcy waiting in the wings. This was not to be the flowing rugby of the French game, but a taut, tactical, attritional battle, which was underlined by the fact that it was almost thirty minutes before the crowd saw any score at all. This, to Irish relief, came from the boot of O'Gara. First blood. It looked like this would be the score at half time but, moments before the break, Ireland conceded a penalty, and England drew level. No one was relaxing now.

The second half began in the same vein, but an early penalty gave O'Gara a chance to double the score. When it struck the post, registering his third miss from four attempts, there was a growing sense that this might not be our day – at least with the boot. But then, as if to mock the non-believer, Brian O'Driscoll, from a full 35 yards, decided to score a drop goal. 6-3. Moments later, he was breaking down the wing, and having chipped threateningly ahead was dangerously obstructed by Armitage. The referee gave the penalty, which O'Driscoll was probably told about after he finished receiving treatment for the painful blow. The crowd were incensed; the tempo of the game rose accordingly, and Ireland pummelled the English line. Phil Vickery of England was sin-binned for disrupting play, and still Ireland were camped on the line. Then O'Driscoll got the ball. And scored. Ireland 11, England 3.

England, of course, did not throw in the towel, and some surging runs brought them back into the danger zone. Continued pressure brought a penalty, and Armitage, the villain of the piece earlier, sent it home. England had been accused of indiscipline in recent games, and this was starting to look like a repeat of form, when Danny Care was sin-binned for a needless charge into the back of Marcus Horan. Ronan O'Gara finally got some luck with a kick, and Ireland led 14-6. This put us out of range from a converted try, but it was no less than we needed, as with moments left to play, a clever chip saw Armitage touch down. Only a point in it now, but the seconds ticked out, and the whistle blew. With this victory, Ireland were to become the only team who remained unbeaten, the only team with the ultimate prize still in sight. Ireland win, 14-13.

A brave England supporter, in Dublin's Temple Bar ahead of the game. Perhaps next year. (Brian Lawless/Sportsfile)

Ireland supporters Shonagh Byrne, left, and Helena Doory, from Roscommon, on their way to the game. (Pat Murphy/Sportsfile)

Ireland supporter Mike Skelley with his father Paul, left, from Liverpool, England, on their way to the game. (Pat Murphy/ Sportsfile)

Captain Brian O'Driscoll, accompanied by mascot Isabel Finlay, leads his side out before the game. (Brendan Moran/Sportsfile)

The England team make their way to the line-up for the national anthems as Ireland's supporters make their presence felt. (Brendan Moran/Sportsfile)

GUINNESS
DIMPLEX
WOLF BLASS
Together is better
Paddy Power
Ulster Bank
POWERADE

Above: England coach Martin Johnson before the game. (David Maher/Sportsfile)

Left: The teams line up for the anthems before the game. (Brendan Moran/Sportsfile)

Above: Nick Kennedy loses possession after being tackled by Brian O'Driscoll. (Brendan Moran/Sportsfile)

Opposite: Paddy Wallace hands off Andrew Sheridan as he is tackled. (Brendan Moran/Sportsfile)

Above: Paul Sackey is tackled by Ronan O'Gara. (Brendan Moran/ Sportsfile)

Left: Donncha O'Callaghan wins possession in the line-out against Steve Borthwick. (Pat Murphy/Sportsfile)

The two Ulstermen, Wallace and Ferris, welcome Toby Flood to Croke Park. (Pat Murphy/Sportsfile)

Tommy Bowe is tackled by Dylan Hartley. (David Maher/Sportsfile)

Brian O'Driscoll is hit by Toby Flood. (Brendan Moran/Sportsfile)

Fly
Emirates

Opposite: Harry Ellis and Rob Kearney prepare to land. (Brendan Moran/Sportsfile)

Above: Rory Best is tackled by Andy Goode, while O'Connell requests the ball. (Brendan Moran/Sportsfile)

Paul O'Connell gets the ball, and gets hit hard by Andrew Sheridan and James Haskell. (Pat Murphy/Sportsfile)

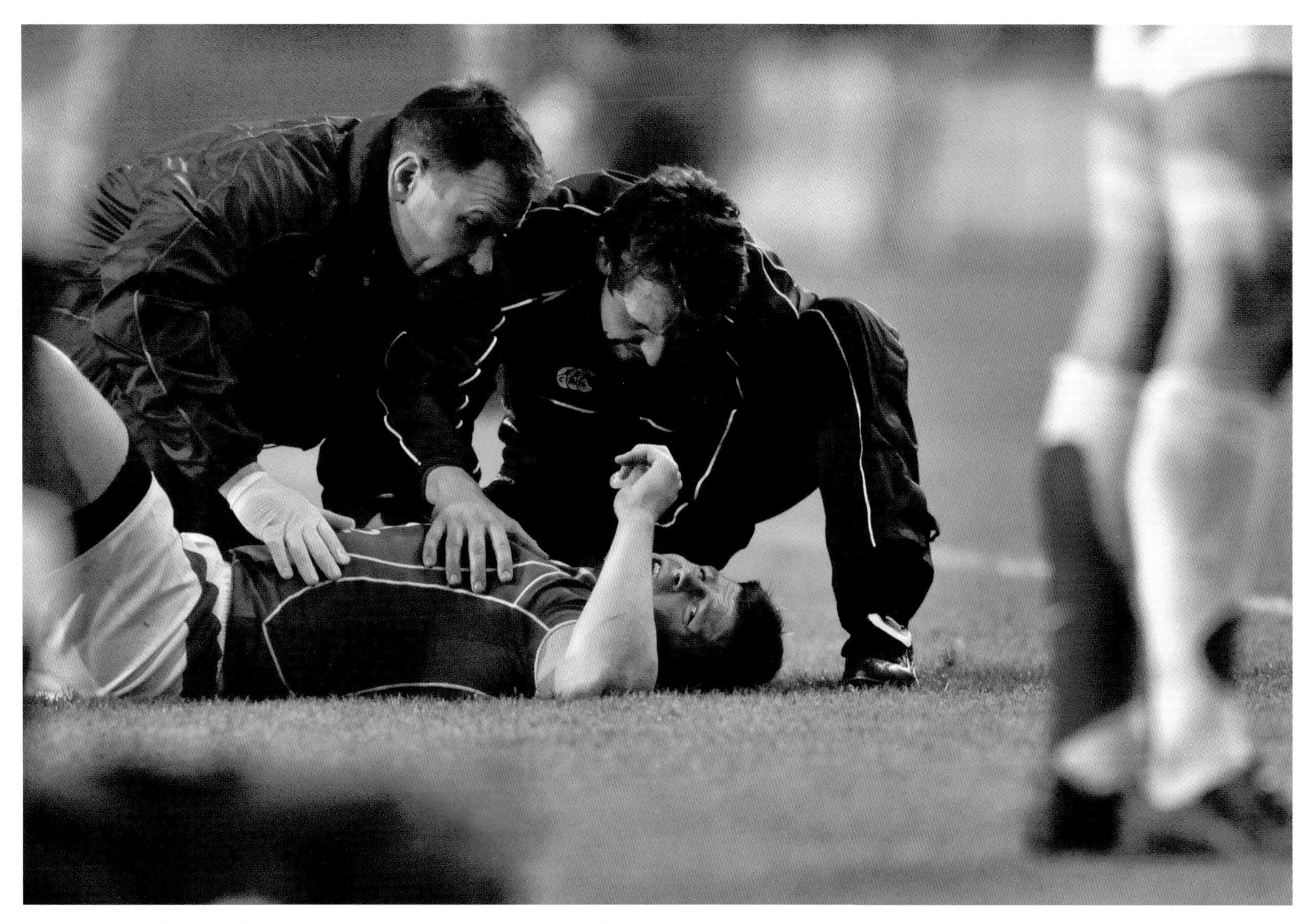

Brian O'Driscoll is attended to by team physio Cameron Steele, right, and team doctor, Dr Michael Webb. (David Maher/Sportsfile)

Above: Brian O'Driscoll shakes off the blow to score his side's first try. (David Maher/Sportsfile)

Opposite: The captain is pleased. (David Maher/Sportsfile)

ENGLAND
O2

Ireland v Scotland

MURRAYFIELD, SATURDAY 14 MAY 2009

And so, to Edinburgh. As always, the sense of occasion in this great city was incredible. With a Grand Slam now becoming a real possibility, there was huge pressure on Ireland, but a proud Scotland were never going to take things lightly at home. Any illusions that this might be the case were quickly dispelled by the imperious Chris Patterson, who stood up to take the first penalty. It has been the best part of half a decade since he missed a kick, and that was not about to change now. Scotland took the lead.

If a kicking gauntlet exists, then it looked as if it was being thrown down today, as O'Gara hit back to level the scores. Patterson put two more away in what was becoming a tight, hard game. With a penalty to bring the scores back to 9-6 to Scotland, O'Gara eclipsed Jonny Wilkinson to become the highest 6 Nations points scorer in history. Chris Patterson showed his regard for this fine achievement by swiftly kicking another score to put Scotland 12-6 ahead. This was not the way it was meant to go. And, with a break down the Irish line, Evans and Godman almost combined to make things a lot worse for Ireland. Godman was clear and away, were it not for a simply sublime covering tackle by O'Driscoll, who made up some 35 yards to get to his man and heave him headlong into touch. Half-time must have come as a relief all round, not least to the fans. Scotland led at this stage, 12-6.

But with a new half came a new Ireland. And who would it be to turn things around but Peter Stringer. Spotting a gap, the veteran scrum half cut free, and set Jamie Heaslip loose to make for the line. Such was the tension that even a professional like Heaslip could seemingly not help celebrating his own score before he had crossed the line, as if he was himself watching the match in a local pub. Who could blame him though, and Ireland regained the lead.

The kicking battle was continuing apace, and after fifty-seven minutes O'Gara upped the stakes by knocking home a perfect drop goal to pull Ireland further away. Ireland now led 19-12, and were finally taking control. With one final penalty, the game was out of the reach of Scottish hands. Ireland win, 22-12. And now it was really on…

An Ireland supporter awaits the start of the game. (Pat Murphy/ Sportsfile)

Scotland reject claims they fielded a weakened team. (Brendan Moran/Sportsfile)

Peter Stringer, left, and Paul O'Connell during 'Ireland's Call'. (Brendan Moran/Sportsfile)

Luke Fitzgerald tackles Scotland's Phil Godman. (Brendan Moran/Sportsfile)

Rob Kearney is upended by Simon Danelli. (Brendan Moran/
Sportsfile)

Committed tackling from Scotland's Phil Godman.
(Brendan Moran/Sportsfile)

1
canterbury

Opposite: Donncha O'Callaghan is caught by Alasdair Dickinson. (Brendan Moran/Sportsfile)

Above: Peter Stringer escapes the outstretched hand of Nathan Hines, in the lead up to Ireland's try. (Brendan Moran/Sportsfile)

RBS

Opposite: Jamie Heaslip celebrates, as he is about to score a try against Scotland. (Brendan Moran/Sportsfile)

Above: Heaslip touches down, despite the challenge of Chris Paterson. (Brendan Moran/Sportsfile)

RBS
RBS

Opposite: Ronan O'Gara kicks the drop goal. (Brendan Moran/Sportsfile)

Above: Jason White, at the final whistle. (Brendan Moran/Sportsfile)

Ireland v Wales

MILLENNIUM STADIUM, SATURDAY 14 MARCH 2009

The last game on the last day of the 6 Nations, and how perfect that was. Wales were the holders of the Grand Slam, Ireland the only team left who could win it. The Championship was still up for grabs, with Wales needing a victory by thirteen points to clinch it. And, of course, the winner of this game would take home the Triple Crown. As if tensions were not high enough, Wales coach Warren Gatland had added fuel to this fire during the week, claiming that if there was one team the Welsh disliked it was Ireland. While it might be a back-handed compliment, that meant nothing on the day, and this incredible stadium was a cauldron of sound and fury.

Ryan Jones, the Welsh captain, did little to calm the situation with an early trip on Ronan O'Gara. Donncha O'Callaghan was immediately on the scene, making his feelings abundantly clear to the big Welshman. When they were finally pulled apart, this enthralling game began again. The tackling was ferocious, hit after hit going in. Ireland had a try disallowed for a forward pass on the line, but after thirty minutes, there was still no score.

O'Gara was clearly being singled out for special treatment by the Welsh, but after thirty-one minutes, it was Stephen Jones who had the first chance at points. And he did not miss. Ireland seemed to have the upper hand after the restart, but again, on thirty-eight minutes, another penalty from Jones put the Welsh further in front. Once again, Ireland were heading in at half-time behind. Declan Kidney's talk at half-time was taking on epic importance in the collective imagination.

And it seemed to work, as Ireland came storming out for the second half, camping themselves on the Welsh line. Again they broke, and again they were held back. But then, stretching through a forest of legs and bodies, an Irish hand looked to touch the ball down, inches over the line. That hand belonged to Brian O'Driscoll. Try. Ireland 7, Wales 6. The Irish crowd went crazy, but when the screaming stopped, the question was whether we could hang on. Ronan O'Gara and Tommy Bowe answered that question some two minutes later. After a deft chip from the out-half, Bowe came hurtling on to the bouncing ball, taking it between two Welsh defenders, and broke for the line. No one was going to catch him. Try. With a simple conversion, Ireland now led 14-6. Could that be the end of the drama? Unlikely.

As 'The Fields of Athenry' echoed around the ground, Wales got another penalty. It was a tough kick, but Jones was on form, and the better you are, the luckier you get; the ball went in off the post. 14-9. On fifty-five minutes, after some barging in the line-out, and there was another penalty for Jones. And another score. 14-12. This was becoming too much to bear.

The game moved into the seventieth minute and it was all Wales now. Tackles were flying in, not least from Stringer, who was throwing himself at everything in red. Still, the Irish line was holding. Seventy-three minutes, and who stepped up again but Stephen Jones. From a ruck the ball was flashed back to him, and as if he was on a silent training ground he calmly struck a drop goal, high and true between the Irish posts. Wales took the lead, 15-14. A deafening silence swept over the Irish crowd, but not for long. Bit by bit, in wave after wave, the Irish drove up the pitch. O'Gara was screaming for the ball, but Stringer held on. It was up to seventy-seven minutes, and then it happened. A hard, swift pass back to the out-half, and another drop-goal attempt. As O'Gara shaped up, the Welsh defenders closed in startlingly fast. Afterwards, the out-half said that they were so close he had to strike the ball upwards rather than straight through as he normally would. Whatever he did, it worked. To ecstatic Irish cries, we were back in front, 17-15. Less than a minute to play. That had to be the end of it. And then the whistle blew. Penalty Wales…

The kick was just inside the Irish half, in roughly the centre of the pitch. It was certainly a tough kick, but within range for Stephen Jones. He could have asked Gavin Henson to take it with his massive range, but as he said after the game, in these situations you have to back yourself. And so he hit it, and as the ball arced towards the posts, time stood still in the stands. As the crowd waited and watched, the ball dipped and dipped and fell just short of the crossbar. There was a moment of silence before the ground erupted. Geordan Murphy, looking shell-shocked, ran towards the crowd and, touching the ball down, kicked it high into the stands, and turned to see all of Ireland realise they had just won the Grand Slam.

Above: Ireland fans, from left, Will McCarthy, Fionan Henry, Eoin Fogarty, from Rathfarnham, Dublin, with Wales fans in Cardiff ahead of the clash with Wales. (Stephen McCarthy/Sportfile)

Right: The Ireland squad stand for the anthems before the final game. (Brendan Moran/Sportfile)

624
625
424
425
LLANDEILO • LLANDOVERY
LLANDUDNO • LLANDYBIE • LLANELLI
LLANELLI WDRS • LLANGEFNI
LLANGENNECH • LLANGWM • LLANHARAN
LLANTWIT MAJOR • LLANYBYDDER
LONDON WELSH • LOUGHOR • MACHEN
MAESTEG • MAESTEG CELTIC • MAESTEG QUINS
MERTHYR • MILFORD HAVEN
webb ellis
SWALEC
SWALEC
O2

Paul O'Connell showed incredible agility and strength in the line-out throughout, and never more so than during the Wales game. (Brendan Moran/Sportfile)

VIKING DIRECT

RBS
O2

Opposite: Tommy Bowe beats the tackle of Ryan Jones…
(Brendan Moran/Sportfile)

Above: Tommy Bowe beats Shane Williams…
(Brendan Moran/Sportfile)

Tommy Bowe goes over for the try!
(Matt Browne/Sportfile)

Jamie Heaslip is tackled by Adam Jones (3) and Alun-Wyn Jones.
(Matt Browne/Sportfile)

Ronan O'Gara kicks that drop-goal, to make the score Wales 15 - Ireland 17.
(Stephen McCarthy/Sportfile)

Ronan O'Gara watches his drop-goal all the way.
(Brendan Moran/Sportfile)

Stephen Jones watches his penalty come up short of the posts, handing Ireland the Grand Slam. (Matt Browne/Sportfile)

BRAINS
SA
CONTINENTAL
WELSH BEER
7:25
CYMRU 15
IWERDDON 17
VILLAGE hotels eatwelshlamb.com
UNDER ARMOUR
UNDER ARMOUR
BRAINS SA GOLD BRAINS SA
RBS
RBS

Rory Best and Paul O'Connell celebrate at the final whistle. (Brendan Moran/Sportfile)

Ireland players, from left, Stephen Ferris, Jamie Heaslip, Tomas O'Leary and Tommy Bowe wait for captain Brian O'Driscoll. (Brendan Moran/Sportfile)

Right: Captain Brian O'Driscoll is congratulated by HRH Prince William before lifting the RBS 6 Nations trophy. (Brendan Moran/ Sportfile)

Brian O'Driscoll lifts the RBS 6 Nations trophy after the game in the company of President Mary McAleese and HRH Prince William. (Brendan Moran/Sportfile)

Paul O'Connell celebrates. (Stephen McCarthy/Sportfile)

Peter Stringer acknowledges the fans.
(Stephen McCarthy/Sportfile)

OLD BRAINS SA GOLD BRAINS SA GOLD

The Ireland team seem pleased with this result. (Matt Browne/Sportfile)

The Homecoming

DAWSON STREET, DUBLIN, SUNDAY 15 MARCH 2009

Sunday, the day after the night before, and thousands of people waited in Dublin Airport for the arrival of their all-conquering heroes. When the team finally emerged, the roars could be heard across the runway. One man, waiting by the team coaches, was told the team would be on a tight schedule when they came out – he responded that he had been waiting for sixty-one years.

The scenes that greeted the Irish team there, however, were dwarfed by the crowds in Dawson Street, where an estimated 17,500 people had descended to scream their respects. From above, the street was a sea of green flags, jerseys, and small children in huge hats. A stage had been erected, and the path to it from the Mansion House was lined with high flags and drummers, beating out a path for the glorious unbeaten. In groups of four or five they emerged to a tumult of noise. All in their team suits, the players looked a little the worse for wear, whether from the championship or the celebrations, it was hard to tell. Donncha O'Callaghan was in exuberant form, signing autographs, posing for photographs, and happily waving like the entertainer he is reputed to be. Paul O'Connell said a few brutally honest words about the team festivities in Cardiff, with a wry grin at the 'single lads', who had made their way into the town to join the Irish crowds after the game. Tommy Bowe did not need too much persuasion to take the microphone, and belted out a couple of verses of 'The Black Velvet Band'. The support that the team had given each other throughout the games did not, seemingly, extend to helping them sing, but Bowe seemed happy enough to go this one alone.

After a few words from the captain, Declan Kidney took the microphone. Surely the most low-key Manager the team could have, it was no surprise that he did not take this opportunity to follow Tommy Bowe in song. Instead, he thanked the fans, the squad, and the 'leaders on the pitch', O'Driscoll, O'Connell, O'Gara and Rory Best. After the applause died down, the players were ushered forward. In this competition the team had reached a few milestones. John 'The Bull' Hayes had become Ireland's most capped player; O'Gara had become the highest points-scorer in the history of the 6 Nations; O'Driscoll was named Player of the Tournament. Many of them had put their names firmly in the hat for the Lions Tour in the summer. But as they raised the championship trophy and the Triple Crown to an explosion of green ticker tape to rapturous cheering from the crowd, it was all about one thing — at long last our Golden Generation had done it. Ireland had won the Grand Slam.

The heaving crowd of fans on Dawson Street. (Stephen McCarthy/ Sportsfile)

A young fan covers his ears from the music as he awaits to greet the Ireland team.(Stephen McCarthy/Sportsfile)

A spectator waves a flag to welcome the players. (Stephen McCarthy/ Sportsfile)

An aeriel view of the large crowd on Dawson Street that gathered to greet the team.
(Stephen McCarthy/Sportsfile)

Donncha O'Callaghan leads players Stephen Ferris, Jamie Heaslip, Mick O'Driscoll and David Wallace as they leave the Mansion House after a civic reception on the team's return. (Stephen McCarthy/ Sportsfile)

Donncha O'Callaghan, left, and Stephen Ferris pictured on the team's homecoming. (Stephen McCarthy/Sportsfile)

Tommy Bowe sings 'The Black Velvet Band' to the crowd.
(Diarmuid Greene/Sportsfile)

Declan Kidney and Brian O'Driscoll lift the Championship trophy as Ronan O'Gara watches on and Paul O'Connell lifts the Triple Crown. (Stephen McCarthy/Sportsfile)

Declan Kidney Brian O'Driscoll lift the RBS 6 Nations Championship trophy, as all Ireland celebrates the Grand Slam. (Stephen McCarthy/Sportsfile)